CSU
Chancellor's
Office

THE CALIFORNIA STATE UNIVERSITY JOURNAL OF SUSTAINABILITY AND CLIMATE CHANGE

Volume 2, 2022

ISBN: 978-1-947112-91-9
Online ISSN: 2771-5582
Print ISSN: 2771-5574

Layout & design by Kiran Dunning

Cover Photos by Kellie Jo Brown, Cal Poly Humboldt, https://www.flickr.com/photos/calpolyhumboldt/

The Press at Cal Poly Humboldt.

CONTENTS

The California State University Journal of Sustainability and Climate Change
2022, Volume 2

Exploring the Use of Trauma Informed Practices in Campus as Lab Programs: Learnings from a Workshop Series

Laurelin Haas (Florida State University), Rachelle L. Haddock (University of Calgary), Joe Fullerton (San Mateo County Community College District)

Abstract

With the intersectional challenges of the climate crisis, the COVID-19 pandemic, and mental health challenges in various forms, empowerment can hold a significant key to mitigating and preventing traumatic experiences at post-secondary institutions. Campus as Lab (CaL) is a growing trend in higher education whereby students, faculty, and staff use experiential learning and applied research projects to advance sustainability on their campuses. It is a unique, empowering learning methodology that can synergistically benefit academic and operational sustainability efforts at post-secondary institutions. In July 2021, a group of professionals who support or lead CaL initiatives gathered to participate in four Summer Series webinars to explore the use of trauma informed practices in CaL programs. This paper provides a high-level overview of the Summer Series webinar structure and explores how participants identified opportunities to use a trauma informed framework for future CaL initiatives. Because of the Summer Series webinars, we believe there is a need for greater familiarity of trauma informed practices on campuses and amongst sustainability staff. Future research could explore the broader application of trauma informed approaches in the various fields of sustainability within post-secondary institutions.

Introduction

Before you begin reading this paper, notice any tension in your body. Roll your shoulders back, and plant your feet gently, yet firmly, on the floor. Take a deep breath. Move your head from side to side. You have just implemented a trauma informed practice.

Trauma informed practices help support those who are coping with trauma by providing strategies for regulating emotions and building resilience. Voluntary slow deep breathing is one example of a trauma informed practice. By decreasing oxygen consumption, heart rate, and blood pressure, deep breathing can lead to a calming effect on the mind and a sense of control of the body (Kim et al., 2014). According to the Agency for Healthcare Research and Quality, an agency within the U.S. Department of Health and Human Services, trauma informed care is "an approach to engaging people with histories of trauma that recognizes the presence of trauma symptoms and acknowledges the role that trauma has played in their lives" (n.d., para. 4). Within this paper, the terms "trauma informed care" and "trauma informed approach" are used interchangeably, while "trauma informed practices" is a term used to refer to the activities one can utilize to honor trauma informed care or approaches.

Research has shown that students in post-secondary institutions are increasingly experiencing a combination of traumatic events, mental health stressors, and climate anxiety (Reyes et al., 2021; Stewart 2021). Today, trauma informed practices are being implemented in colleges and universities to create supportive environments that promote student resilience and success (Barros-Lane et al., 2021; Davidson, 2017). This paper explores the novel use of trauma informed practices in the context of post- secondary Campus as Lab (CaL) programs.

CaL may be referred to as Campus as a Living Lab, Campus as a Learning Lab, Living Classroom, Applied Learning in Sustainability, or other terms. CaL activity can focus on a diversity of aims including cultivating sustainability leaders, demonstrating scalable sustainability impact, and harnessing academic research expertise toward global sustainability problem-solving (Haddock & Savage, 2020). For the purposes of their sustainability benchmarking system, the Association for the Advancement of Sustainability in Higher Education (AASHE) (2021) defines CaL through institutions that:

> [U]tilize their infrastructure and operations as living environments for multidisciplinary learning and applied research that advances sustainability on campus. Students that actively participate in making their campuses more sustainable are well prepared to continue that work in their careers and communities after graduation. (p. 1)

CaL is a unique, empowering learning methodology that can synergistically benefit academic and operational sustainability efforts at post-secondary institutions. There is a growing body of research that suggests engaging students in experiential opportunities like CaL can help combat the negative effects of climate anxiety (Aruta & Simon, 2022; Bentz, 2020; Buchanan et al., 2021; Kelly et al., 2021). However, little research has been completed that specifically addresses CaL and its potential for reducing the effects of multiple stressors.

With the intersecting challenges of the climate crisis, the COVID-19 pandemic, and mental health challenges in various forms, empowerment can hold a significant key to mitigating and preventing traumatic experiences. We posit that CaL's applied theory-to-practice framework can empower students to have a hand in real-world climate and sustainability solutions and can serve as an effective way to integrate trauma informed care into higher education.

The purpose of this paper is to:

1. Provide a high-level overview of trauma informed care and its relevance to CaL;
2. Share the learnings from the four 2021 Summer Series webinars to explain how partnership, collaboration, and opportunities for engagement can help build individual and organizational resilience to trauma; and,
3. Suggest opportunities for future research to explore the broader applications of trauma informed approaches within sustainability initiatives at post-secondary institutions.

The target audience for this paper is sustainability professionals, both faculty and support staff, who lead and support CaL activities. The literature cited within this paper on trauma informed practices in higher education does not differentiate between types of higher education professionals. In these instances, the authors use the term higher education professionals to convey the research accurately.

CaL and Trauma Informed Care

CaL is a growing trend at post-secondary institutions whereby students, faculty, and staff use experiential learning and applied research projects to advance sustainability on their campuses (Haddock & Savage, 2020). Multiple scholars have emphasized the need for these approaches in environmental education, noting that direct experience in tackling campus and community projects can help develop environmentally responsible citizens (Chawla & Cushing, 2007; Hungerford & Volk, 1990).

Need for New Approaches amidst a Convergence of Stressors

The impetus to explore trauma informed care as a framework for CaL projects stemmed from the growing acknowledgement of the negative impacts of trauma, mental health stressors, such as the COVID-19 pandemic, and the ongoing climate crisis on the students, faculty, and staff who design, implement, and participate in sustainability change efforts and CaL initiatives.

According to the Substance Abuse and Mental Health Services Administration, a branch of the U.S. Department of Health and Human Services, trauma can be defined broadly as:

> The experiences that cause intense physical and psychological stress reactions. [Trauma] can refer to a single event, multiple events, or a set of circumstances that are experienced by an individual as physically and emotionally harmful or threatening and that have lasting adverse effects on the individual's physical, social, emotional, or spiritual well-being. (2014b, p. xix)

Today, trauma is widespread. The Centers for Disease Control and Prevention (2019) indicate that adverse childhood experiences, or ACEs, are potentially traumatic events that occur in childhood (zero-17 years). Further, ACEs can include 1) experiencing violence, abuse, or neglect; 2) witnessing violence in the home or community; and 3) having a family member attempt or die by suicide. ACEs are quite common, even among a middle-class population, and there is a powerful, persistent correlation between the more ACEs experienced and the greater the chance of poor outcomes later in life (Center on the Developing Child - Harvard University, 2022). The Centers for Disease Control and Prevention (2019) reported that 61% of adults have had at least one adverse childhood experience (ACE), and 16% have had four or more types of ACEs.

In a 2019 study, 70% of freshmen entering college reported experiencing at least one potentially traumatic event, and 34.4% of the trauma-exposed individuals met criteria for probable post-traumatic stress disorder (Cusack et al., 2019). Particular sub-groups of students, including first generation college students, Indigenous students, LGBTQ+ students, and student veterans, may have additional mental health concerns, which could make them more vulnerable to experiencing symptoms of PTSD, depression and anxiety (Davidson, 2017; House et al., 2020; Morissette et al., 2021; Travers et al., 2020). Refugee students may have experienced trauma because of war, civil unrest and family disunity (Erisman & Looney, 2007). Students entering post-secondary institutions after aging-out of the foster care system may also have experienced trauma; rates of post-traumatic stress disorder were twice as high for youth aging-out of care compared to the American war veteran population (Gomez et al., 2015).

The threat of climate change can lead students to experience apocalyptic fears of annihilation and extinction as well as pre-traumatic stress (Dodds, 2021; Panu, 2020). A recent study revealed that students felt "overwhelmed," "angry," and "ashamed" when asked how they felt about climate change (Hiser & Lynch, 2021). Student responses to climate anxiety have included outright denial, manic defense behaviors, and burnout (Dodds, 2021).The mental health and climate crises are pervasive, compounding issues that require attention and investment (Romeu, 2021).

The COVID-19 pandemic has increased student stress, particularly related to academic workloads, separation from school, and fears of contagion (Yang et al., 2021). The COVID-19 pandemic exacerbates the impact of stress and trauma on individuals, organizations, and institutions (Bridgland et al., 2021). These and other types of collective trauma require focus and action by higher education institutions; indeed, the "...combination of medical, economic, racial and climate-based catastrophes highlights the need for attention to the meaning and implications of cumulative, compounding trauma exposure" (Silver et al., 2020).

Prolonged student exposure to the COVID-19 pandemic and the climate crisis represents a new challenge for higher education support staff and faculty members. Sustainability professionals at post- secondary institutions are positioned to address these interrelated challenges. Those who work within sustainability can directly address issues relating to climate change and involve students in local solutions to overwhelming global challenges. Sustainability professionals can re-focus feelings of hopelessness and apathy and transform them into action by facilitating experiential learning opportunities, such as CaL (Roysen & Cruz, 2020).

Using trauma informed approaches is one way that higher education professionals can better support students who have been impacted by trauma. For example, a 2021 study found that when trauma informed care interventions were incorporated into higher education COVID-19 response, the interventions fostered a sense of safety, encouraged students' empowerment, and created opportunities for connection and support (Barros-Lane et al., 2021). However, there is a dearth of previous studies on the use of trauma informed practices to advance sustainability in higher education. Trauma informed approaches have been discussed in relation to living labs in the context of child protection and domestic and family violence sectors (Wendt et al., 2021); however, a keyword search of "living lab" and "trauma" yielded no results relevant to higher education.

Climate anxiety can also negatively affect the well-being of higher education staff and faculty members (Gilford et al., 2019). Therefore, sustainability professionals should have cursory knowledge of the concepts, theories, and practices of trauma informed care so that they can empower others while remaining responsive to their own mental health needs. Sustainability professionals should look to these practices when engaging in change-making efforts, especially those that are not guaranteed success, to maintain healthy boundaries, develop realistic expectations, and facilitate positive experiences for all involved.

Trauma informed care, trauma informed principles, and trauma informed practices

Trauma informed care operates according to six basic trauma informed principles, which are outlined in Table 1. These are safety, trustworthiness and transparency, collaboration and mutuality, empowerment and choice, peer support and mutual self-help, and cultural, historical, and gender issues (National Child Traumatic Stress Network, n.d.).

Table 1
The six trauma informed principles and brief descriptions

Principles of Trauma Informed Care	Description
Safety	Physical and psychological safety
Trustworthiness and transparency	Clear tasks, consistency, and appropriate boundaries
Collaboration and mutuality	Partnering and leveling of power differences
Empowerment, voice, and choice	Recognizing, building, and validating skills; strengthening the experience of choice; recognizing that every person requires an individualized approach
Peer support and mutual self- help	Valuing lived expertise
Cultural, historical, and gender issues	Moving past cultural stereotypes and biases

Note: Reprinted from "SAMHSA's Concept of Trauma and Guidance for a Trauma-Informed Approach", by Substance Abuse and Mental Health Services Administration, 2014, HHS Publication No. (SMA) 14- 4884, 11.

A trauma informed approach and the principles outlined above can be instituted at any level and by any type of organization (Raja et al., 2015). According to Menschner and Maul (2016), a trauma informed approach shifts the focus from "What's wrong with you?" to "What happened to you?" by:

- Realizing the prevalence of traumatic events and the widespread impact of trauma;
- Recognizing the signs and symptoms of trauma;
- Responding by integrating knowledge about trauma into policies, procedures, and practices; and
- Seeking to actively resist re-traumatization. (p. 2)

Structure and Learnings from the 2021 Summer Series Webinars

The 2021 Summer Series webinars were a set of four consecutive online workshops. Sustainability professionals were the primary audience for the series, and each workshop featured a different topic related to overcoming the challenges of a COVID-19-impacted environment. Participants included both faculty members and support staff who identify as sustainability professionals at higher education institutions in both Canada and the United States. The majority of the participants were members of the CaL Community of Practice (CoP). The aim of the CaL CoP is to advance experiential learning and applied research at university and college campuses to address sustainability challenges. Founded in 2016, the CoP provides a foundation for experimenting cross-institutionally on novel approaches to engaging in CaL efforts.

Structure of the 2021 Summer Series Webinars

Over the course of four webinars, organizers used a progressive narrative arc (Figure 1.) to help participants develop a shared terminology, define challenges in their institutional contexts, learn from best practices, and explore potential partnerships. Throughout the webinars, participants learned about trauma on a variety of scales, moving from personal impacts of the COVID-19 pandemic to institutional actions and partnerships that can help address the ongoing trauma of climate change.

The main goal of the Summer Series webinars was to demonstrate to participants how trauma informed practices could be integrated into their work. To achieve this goal, organizers designed the webinars to incorporate and model the principles of trauma informed care in the following ways:

- Safety: Organizers established workshop norms, including the norm of confidentiality, to provide emotional safety for participants.
- Trustworthiness: Each webinar began with a clear outline and stated objectives to develop trust.
- Collaboration: Organizers shared power by integrating participant feedback into future webinar designs.
- Choice: Whenever possible, participants were provided with multiple ways to move forward in a webinar. For example, organizers gave participants many opportunities to take a break, share feedback in the chat, or work in small groups. Organizers also prioritized tangible skills, such as applications of the principles of trauma informed care, to empower participants to utilize knowledge gained from the webinars.
- Peer Support: Working in small breakout rooms, participants had an opportunity to engage peer-to-peer to generate ideas and develop trauma informed approaches to sustainability challenges.

Figure 1
The Summer Series Webinars narrative arc transitioned from exploring personal trauma to addressing organizational opportunities to address trauma.

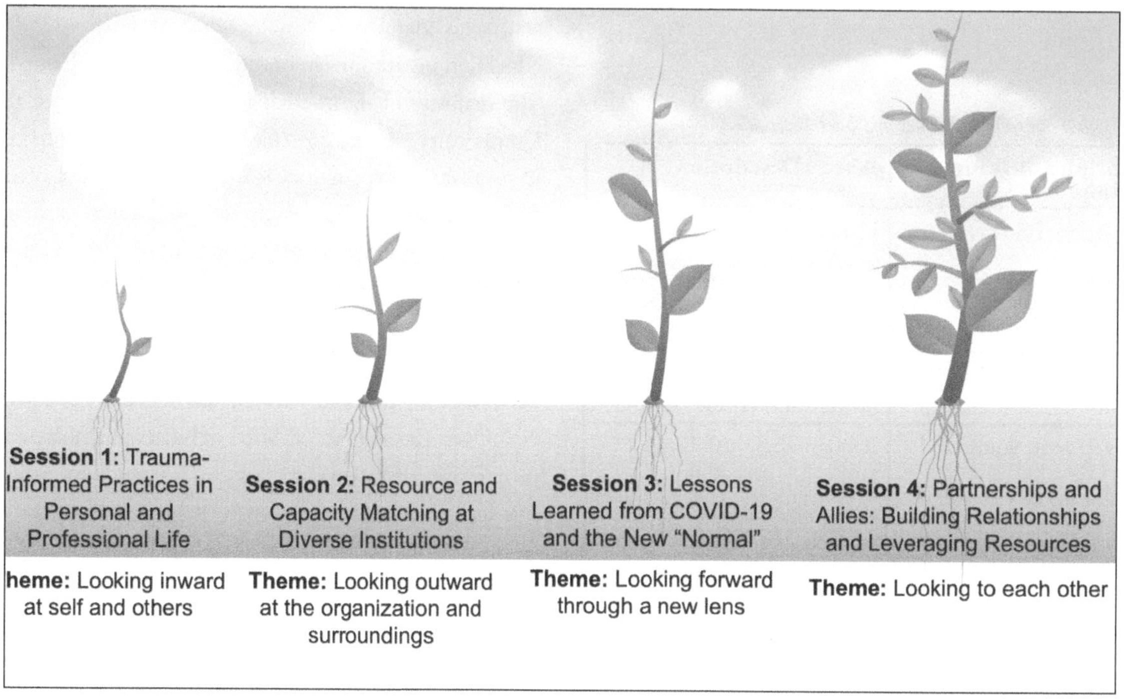

- Cultural, Historical, and Gender Issues: Organizers began each webinar with a land acknowledgement to bring issues of justice, equity, diversity, and inclusion (JEDI) to the forefront of each workshop.

Learnings from the 2021 Summer Series Webinars

The four Summer Series webinars were held in July 2021. There were 26 unique participants over the course of the series. A full list of webinar objectives is provided in Table 2 and the following section details each webinar's structure and learnings.

Webinar One: Trauma Informed Practices in Personal and Professional Life.

In webinar one, a county Clinical Services Manager specializing in trauma informed systems advancement led 24 participants through a presentation on trauma informed practices and their application in different environments.

The guest speaker connected trauma informed practices to sustainability through a discussion on the impacts of prolonged uncertainty and stress, which can be caused by climate anxiety and the specific risk factors of college-age students. Finally, participants explored practices that they could apply in interactions in the workplace and in their personal lives. For example, as a self-care practice, participants were challenged to identify: 1) "What are your flags that something is wrong?" 2) "What soothes you?" and 3) "Who are you going to talk to?" Given that this webinar was focused on establishing a common understanding of the basics of trauma informed practices, there are no specific outcomes to share.

Webinar Two: Resource and Capacity Matching at Diverse Institutions.

In webinar two, 21 participants reviewed case studies of resource and capacity matching at post-secondary institutions. Guest speakers from an environmental education nonprofit led participants through a series of small group discussions. These conversations focused on both the changing institutional environment after the COVID-19 pandemic and how professionals could leverage changes to advance sustainability through CaL initiatives.

The small group discussions explored how sustainability professionals could take advantage of current risks to introduce new, resilient responses to challenges. Participants collaborated in small groups to brainstorm around the topics of "Focus," "Vision," and "Change," which are the precursors to action, and shared their ideas using a Google Jamboard.

"Focus" challenged participants to examine the way the COVID-19 pandemic has recalibrated institutional priorities, and participants noted changing attitudes towards equity, diversity, and inclusion, as well as change and uncertainty. "Vision" asked participants to examine what the higher education landscape might look like in the future. Participants saw opportunities in environmental justice and human-centered programs, an alignment of institutional priorities with sustainability, and utilizing new tools for engagement, such as online collaboration. Finally, "change" asked participants to reflect on skills or resources that were needed in order for their desired visions to take place. Participants identified momentum, intentionality, and inclusivity as important values moving forward.

Table 2
2021 Summer Series webinar titles and objectives

Webinar	Objectives
Webinar One: Trauma Informed Practices in Personal and Professional Life	Participants will understand the principles and concepts of trauma informed practices and explore how they can be applied to their personal and professional life.
Webinar Two: Resource and Capacity Matching at Diverse Institutions	Participants will explore the ways COVID-19 has recalibrated institutional priorities. Participants will understand the concepts of resource and capacity matching and how they can be used to overcome barriers.
Webinar Three: Lessons Learned from the COVID-19 Pandemic and the New "Normal"	Participants will share lessons learned from the COVID-19 pandemic and explore ways they can preserve and cultivate the positive aspects of the new "normal" and practice self-care.
Webinar Four: Partnerships and Allies: Building Relationships and Leveraging Resources	Participants will explore how to build and strengthen relationships and share resources across institutions.

Webinar Three: Lessons Learned from the COVID-19 Pandemic and the New "Normal."

In webinar three, 21 participants shared lessons learned from the COVID-19 pandemic and explored best practices moving forward. During the webinar, participants reflected on the ways the COVID-19 pandemic had changed cultural norms and workplace expectations, dividing their experiences into practices they would like to "keep" and practices they would prefer to "throw away."

Participants shared that they would like to "keep" the flexibility, partnerships, and collaboration that they experienced during the COVID-19 pandemic. They also emphasized the importance of empathy and the new focus on justice, equity, diversity. and inclusion. However, participants noted that they would prefer to "throw away" anxiety and burnout from the pandemic. They also shared that during the COVID-19 pandemic, sustainability was often deprioritized, and progress was hampered by uncertainty and lack of guidance.

Next, participants built upon the ideas generated in webinar two and took initial steps to operationalize them. Working in small groups, participants generated action items for four different scenarios, including The Activist Student, New Tools for Engagement, a Human-Centered Approach, and Alignment with Organizational Priorities. Participants applied a trauma informed lens to each case study, which served as an opportunity to apply their learnings across the webinars.

Specifically, in the Activist Student Scenario, participants were asked, "How can we protect and promote student interests without placing the burden of action solely upon them?" Participants noted that it would be important to prevent overtaxing students while also supporting student-driven initiatives that focus on issues students care about. Applying a trauma informed lens to their proposed solutions, participants noted that they could share power by collaborating with different equity groups, offer students choice in their roles and tasks, and promote student-driven peer mentorship.

In the New Tools for Engagement scenario, participants were asked, "How do we ensure that we use new tools of engagement in an equitable and inclusive way?" Participants suggested using virtual platforms to reach wider audiences, keeping information in a virtual space, and working with stakeholders to tailor programs to their needs and wants. Taking a trauma informed approach, participants noted they could hold events in spaces that are accessible to everyone, keep virtual events limited to a set amount of time to reduce burnout, and set clear outcomes and expectations for each event.

In the Human-Centered Approach scenario, participants were asked, "What steps and actions can sustainability teams/ leaders take to include and encourage diversity in sustainability

Table 3
Actions and trauma informed approaches for four scenarios

Scenario	Actions	Trauma Informed Approaches
The Activist Student: "How can we protect and promote student interests without placing the burden of action solely upon them?"	1. Engage freshmen as quickly as possible with sustainability initiatives 2. Supporting intersectionality and connecting with equity groups 3. Connecting with alumni 4. Trying to prevent overtaxing students 5. Student-driven and focusing on what they care about 6. Peer mentorship 7. Connecting with faculty advisors	1. Spreading out tasks to offer students choice in their roles 2. Sharing of power by collaborating with different equity groups 3. Include people with differing abilities through Zoom meetings and closed captioning 4. Provide a balanced approach to discussing large issues of sustainability 5. Student-driven peer mentorship needs to be empowering 6. Empowering everyone and taking into account histories
New Tools for Engagement: "How do we ensure that we use new tools of engagement in an equitable and inclusive way?"	1. Utilize virtual platforms to reach wider audience (e.g., live streaming events) 2. Keeping information in a digital space for record keeping, transparency, and accessibility 3. Working with stakeholders to tailor programs to their needs and wants 4. Tie health and wellbeing into sustainability	1. Keep bio/wellness breaks 2. Asynchronous options for flexibility 3. Choosing spaces that are Americans with Disabilities Act-approved and accessible to everyone 4. Setting clear outcomes and expectations, offering the opportunity before the event to give opinion on what should be included 5. Choosing food options for everyone, keeping in mind cultural and religious backgrounds 6. Gender-neutral restrooms 7. Keeping virtual events limited to a certain amount of time to avoid Zoom burnout
Justice, Equity, Diversity, and Inclusion (JEDI) and Taking a Human-Centered Approach: "What steps and actions can sustainability teams/ leaders take to include and encourage diversity in sustainablility efforts?"	1. Expand from environmental sustainability to include cultural/social sustainability 2. Collaborating with student organizations/affinity groups on campus 3. Provide space/platform for diverse perspectives on sustainability 4. Naming JEDI as a priority 5. Film screenings 6. Amplifying and elevate other voices: not assuming we are the experts 7. Incorporating JEDI & partnership building into onboarding practices	1. Continuing to check-in; acknowledging ongoing crises 2. Approaching these issues with humility 3. Referring to more knowledgeable/specific resources 4. Setting clear expectations for projects/events, naming how one can go about interrupting/taking a pause/ excusing oneself as needed 5. Using shared trauma (COVID- 19/natural disasters) as a shared experience from which to empathize with more culturally/community-specific trauma
Alignment with Organizational Goals: "How do we ensure that we use new tools of engagement in an equitable and inclusive way?"	1. Get students involved - student action is one of the most effective ways to get university leadership onboard 2. Take advantage of restructuring/strategic planning processes 3. Use the celebration of being back on campus to encourage CaL exercises 4. Collaborate with diverse groups on campus for fresh perspectives and input - more voices equals more change 5. Influential to have a member of facilities/academia/ office of sustainability/etc. working together to align and complete sustainability projects 6. Universities motivated by United Nations Sustainable Development Goals and rankings	1. Providing multiple choices of sustainability projects 2. Provide project ideas with phasing/timeline 3. Safe space: health is a primary concern for administration; show CaL/sustainability efforts for their connection to mental and physical health 4. Trustworthiness can come from setting boundaries of timeliness and clear tasks and goals

efforts?" Participants suggested amplifying and elevating other voices, providing a platform for diverse perspectives on sustainability, and expanding to include cultural and social sustainability topics. The participants suggested that certain trauma informed approaches could enhance these efforts, for example, continuing to check-in with their communities, making room for conversations around inequalities, and referring to other resources.

In the Alignment with Organizational Priorities scenario, participants were asked, "How do we surface inconvenient truths and ask for change while the organization and the people within it are still enduring trauma?" Participants suggested getting students involved, promoting CaL initiatives, and taking advantage of restructuring or strategic planning processes. Using a trauma informed approach, participants noted that it was important to set boundaries, provide project ideas with specific timelines, and provide multiple choices in sustainability projects. Table 3 provides a summary of the actions and trauma informed approaches generated for each scenario by participants.

Webinar Four: Partnerships and Allies - Building Relationships and Leveraging Resources

In webinar four, 16 participants explored ways that their institutions could engage with external partners to work together on climate solutions. The webinar emphasized the importance of trauma informed approaches in CaL initiatives, specifically the sharing of power through collaboration with different equity groups. A guest speaker from a climate solutions nonprofit highlighted resources and case studies related to successful sustainability interventions in a variety of fields. Participants were challenged to reflect individually on community resources, professional skills, and institutional needs in order to identify areas for future sustainability work. Finally, participants were invited to contemplate actions and potential collaborations they would pursue following the conclusion of the Summer Series webinars.

Key Takeaways

After the conclusion of the 2021 Summer Series webinars, the authors arrived at several key takeaways. Engagement in the webinars represented a substantial commitment of time and energy for participants and demonstrated that there was an appetite for trauma informed practices amongst sustainability professionals. It is important to continue to support this interest in trauma informed practices by using support networks, such as the CaL CoP or AASHE, in which sustainability professionals can share best practices, challenges, and learnings.

Participants were already using some trauma informed practices prior to the Summer Series webinars although they may not have identified them as such. The practice of creating a safe and supportive learning environment is one example of a trauma-informed practice that is employed by sustainability professionals. Other simple things that can help alleviate fear, anxiety, and stress over discussing and acting on issues as pervasive and complex as the climate crisis include: connecting students to the academic community; providing students with opportunities to practice their skills, embrace teamwork, and participate in shared leadership; and anticipating and adapting to the changing needs of students and the community (Hoch et al., 2015).

The 2021 Summer Series webinars revealed a lack of resources related to trauma informed practices and their application specific to experiential learning and sustainability in higher education. There are many resources broadly related to trauma informed care in colleges and universities, however the authors of this paper were unable to find specific academic papers, applied examples, or case studies of trauma informed practices being applied in CaL-like scenarios.

There is a need for more training provided by subject matter experts in trauma informed care for sustainability professionals focused on implementing trauma informed practices in CaL. The 2021 Summer Series webinars introduced the topic of trauma informed care; however, the organizers were not experts in trauma informed care and could not provide a complete and comprehensive education on the topic. It can be challenging for sustainability professionals, and likely higher education professionals in general, to apply or teach content if they are not experts. Further, additional training in trauma informed care for sustainability professionals is necessary to ensure implementation of trauma informed practices in a way that does not unintentionally re-traumatize or exclude some students.

Finally, the 2021 Summer Series webinars highlighted the cumulative benefits of trauma informed practices. By more intentionally implementing these practices, sustainability professionals have the potential to have a greater positive impact on student participants in CaL programs through creating safe environments in the face of growing climate uncertainty, student trauma, and other ongoing mental health stressors.

Opportunities and Limitations for Future Research

There is a growing body of scholarship on the opportunities for sustainability-focused experiential learning (Favaloro et al., 2019; Gunnels et al., 2021; Rogers et al., 2021;

Rukspollmuang et al., 2022). However, there is little existing research demonstrating how CaL projects can harness trauma informed practices to relieve student stress and anxiety.

Summer Series webinar participants indicated a need for further study on trauma informed care related to climate disasters and requested more examples of how sustainability professionals could interact with students related to climate anxiety. Addressing these knowledge gaps could help sustainability professionals prepare the next generation of sustainability leaders while supporting their own mental health. Drawing a clearer connection between the empowerment of individuals, including students, staff, faculty, and community members, and CaL initiatives is a critical next step.

There is an opportunity to study the effectiveness of the 2021 Summer Series webinars as an instructional method. Future research could track the longitudinal effects of participation in the workshop series. Researchers could explore topics such as: 1) did the webinar series change participants' approach to their work over time; and 2) what is the impact of using trauma informed approaches for CaL initiatives from various perspectives, including students, staff, and faculty members? Additional research could explore the potential to scale-up trauma informed practices in sustainability across post-secondary institutions through additional offerings of a similar workshop.

There are several limitations to addressing the effectiveness of a series of webinars as an instructional method. One limitation is future participation in online workshops. Having a larger group of participants can enable perspectives that are more diverse and enrich webinar learnings. However, the 2021 Summer Series webinars participants acknowledged that interest in online webinars might decrease as higher education institutions shift back to in-person experiences at this point in the COVID- 19 pandemic.

A further limitation is participant engagement after the series of webinars. In order to assess the 2021 Summer Series webinars' effectiveness, it would be ideal to check if participants have utilized their learnings. However, it can be difficult to maintain communication with participants following the conclusion of an online educational series. After the 2021 Summer Series webinars, there was no follow-up to check if participants had incorporated trauma informed practices into their work. As a result, the authors cannot stipulate if the workshop series enabled sustainability professionals to use the practices or if the professionals found the practices to be helpful within the context of advancing sustainability in higher education.

Conclusion

The 2021 Summer Series webinars provided an opportunity to explore ways of incorporating trauma informed care into CaL efforts to address the intersecting challenges of climate change, mental health, and trauma. Beyond what participants individually gained from each of the four webinars, the collaborative project demonstrated that there is an appetite for trauma informed care training for sustainability professionals and a need for further research on this topic. This paper highlights the potential for this approach and provides a replicable framework that can be utilized by others to explore the application of trauma informed approaches to sustainability. While this paper focuses on sustainability professionals, the findings presented here are applicable to higher education professionals in general. The efforts described here represent a first attempt to explore this intersection and highlights avenues that can be expanded upon through future work.

CaL programs should empower students in a safe environment (Rogers et al., 2021). Furthermore, safety in experiential learning should include both physical and psychological health (Pickens & Tschopp, 2017). To enhance participant safety in programs like CaL, sustainability professionals should be familiar with trauma informed practices and understand how to apply them in their work. Furthermore, sustainability professionals should view these practices as beneficial to other goals, such as behavior change. Research suggests that collaborative approaches, including trauma informed care, make it more likely that participants will accept discussions around behavior change and active engagement (Raja et al., 2015). Because many sustainability challenges involve behavior change, sustainability professionals may be more motivated to become early adopters and promoters of trauma informed practices.

As sustainability professionals continue to explore how these techniques can be applied in their daily work, it is important that they be given sufficient resources to ensure sensitive implementation of trauma informed approaches. Resources could include funding for appropriate training sessions, access to materials created by subject matter experts, and the staff time required to attend training sessions and review materials. It is also important that sustainability professionals engage with experts in trauma informed care to share best practices and discuss challenges. Organizations such as the CaL CoP and AASHE could provide a forum for sustainability professionals to continue to collaborate on this topic through work groups, webinars, and conference sessions.

Utilizing trauma informed approaches is an important way that sustainability professionals can work to ensure that CaL participants remain safe and are not re-traumatized through their campus experiences. While the topic of trauma informed care might seem daunting, sustainability professionals are encouraged to approach the topic one opportunity or—empathetic inquiry—at a time.

Acknowledgements

The authors wish to acknowledge the contributions of the numerous speakers who presented as subject matter experts during the Summer Series webinars and who invested their time and expertise in co-designing the webinars: Mary Fullerton, Dr. Elizabeth Bagley, Dr. Jake Pollack, and Tyler Valdes. The authors also wish to thank the participants of the Summer Series webinars who shared their time, expertise and ideas with us during and after the workshop series. All participants were notified via email about a possible publication stemming from the workshop series. The authors express their gratitude to a number of reviewers who greatly improved the clarity and cohesion of this paper, including, Dr. Carrie Ferraro, Evan Christo, Fatima Abdul-Khabir, Madeline Willett, and Travis Milnes. Finally, the authors wish to thank the two anonymous reviewers whose feedback greatly improved the quality of this paper.

References

Agency for Healthcare Research and Quality. (n.d.). *Trauma Informed Care.* Retrieved September 25, 2022 from https://www.ahrq.gov/ncepcr/tools/healthier-pregnancy/fact-sheets/trauma.html

Aruta, J.J. & Simon, P.D. (2022). Addressing climate anxiety among young people in the Philippines. *The Lancet Planetary Health,* 6(2), E81-E82. https://doi.org/10.1016/S2542-5196(22)00010-9

Association for the Advancement of Higher Education (AASHE). (2021). STARS 2.2 Technical Manual - AC 8: Campus as a Living Laboratory. https://drive.google.com/file/d/1JFAXiMPjLOB1C0dJKK7C8R6vfo-Ch5J76/view

Barros-Lane, L., Smith, D.S., McCarty, D., Perez, S., & Sirrianni, L. (2021). Assessing a Trauma-Informed Approach to the COVID-19 Pandemic in Higher Education: A Mixed Methods Study. *Journal of Social Work Education,* 57, S66-S81. https://doi.org/10.1080/10437797.2021.1939825

Bentz, J. (2020). Learning about climate change in, with and through art. *Climatic Change,* 162, 1595- 1612. https://doi.org/10.1007/s10584-020-02804-4

Bridgland, V., Moeck, E. K., Green, D. M., Swain, T. L., Nayda, D. M., Matson, L. A., Hutchison, N. P., & Takarangi, M. (2021). Why the COVID-19 pandemic is a traumatic stressor. *PloS one,* 16(1), e0240146. https://doi.org/10.1371/journal.pone.0240146

Buchanan, J., Pressick-Kilborn, J., & Fergusson, J. (2021). Naturally enough? Children, climate anxiety and the importance of hope. *The Social Educator,* 39(3), 17-31. http://hdl.handle.net/10453/152227

Centers for Disease Control and Prevention. (2019). *Adverse Childhood Experiences (ACEs).* Retrieved April 7, 2022 from https://www.cdc.gov/vitalsigns/aces/index.html

Center on the Developing Child - Harvard University. (n.d.) *ACEs and Toxic Stress: Frequently Asked Questions.* Retrieved September 25, 2022 from https://developingchild.harvard.edu/resources/aces-and-toxic-stress-frequently-asked-questions/

Chawla, L. & Cushing, D.F. (2007). Education for strategic environmental behavior. *Environmental Education Research,* 13(4), 437-452. https://doi.org/10.1080/13504620701581539

Cusack, S.E., Hicks, T.A., Bourdon, J. Sheerin, C.M, Overstreet, C.M., Kendler, K.S., Dick, D.M., & Amstadter, A.B. (2019). Prevalence and predictors of PTSD among a college sample. *Journal of American College Health,* 67(2), 123-131. https://doi.org/10.1080/07448481.2018.1462824.

Davidson, S. (2017). *Trauma-Informed Practices for Postsecondary Education: A Guide.* Education Northwest. https://educationnorthwest.org/sites/default/files/resources/trauma-informed-practices-postsecondary-508.pdf

Dodds, J. (2021). The psychology of climate anxiety. *BJPsych Bulletin,* 45, 222-226. https://doi:10.1192/bjb.2021.18.

Erisman, W., & Looney, S. (2007). *Opening the Door to the American Dream: Increasing Higher Education Access and Success for Immigrants.* Institute for Higher Education Policy. https://files.eric.ed.gov/fulltext/ED497030.pdf

Favaloro, T., Ball, T, & Lipschutz, D. (2019). Mind the Gap! Developing the Campus as a Living Lab for Student Experiential Learning in Sustainability. In W. Leal Filho & U. Bardi (Eds.), *Sustainability on University Campuses: Learning, Skills Building and Best Practices* (pp. 91-113). Springer. https://doi.org/10.1007/978-3-030-15864-4_7.

Gillford, D., Moser, S., DePodwin, B., & Watson, S. (2019, December 6). The Emotional Toll of Climate Change on Science Professionals. EOS. https://eos.org/features/the-emotional-toll-of-climate-change-on-science-professionals

Goldenson, R.P., Avery, L.L., Gill, R.R., & Durfee, S.M. (2021, July 4). The Virtual Homeroom: Utility and Benefits of Small Group Online Learning in the COVID-19 Era. *Current Problems in Diagnostic Radiology,* 51(2), 152-154. https://doi.org/10.1067/j.cpradiol.2021.06.012.

Gomez, R.J., Ryan, T.N., Norton, C.L., Jones, C., & Galan-Cisneros, P. (2015). Perceptions of Learned Helplessness Among Emerging Adults Aging-out of Foster Care. *Child Adolescent Social Work Journal,* 32:507–516.

Gunnels, C.W., Abercrombie, M.I., Bovard, B.D., Croshaw, D.A., Drummond, J., Everham, E.M., Hancock, T., Herman, J., Jackueux, M., Lefevre, K.L., Marcolini, J.P., McConnell, V., Metcalf, M., Sanchez, I.L., Walsh-Haney, H., & Voytek, M. (2021). Campus as a Living Laboratory: Conservation Areas that Create a Community Who then Ensure Its Sustainability. In K. Leone, S. Komisar, & E.M. Everham III (Eds.) *Making the Sustainable University* (pp.133-149). Springer. https://doi.org/10.1007/978- 981-33-4477-8_9.

Haddock, R. & Savage, C. (2020). Campus as Lab: Using Post-secondary Campuses as Petri Dishes for Advancing the Sustainable Development Goals. In E. Sengupta, P. Blessinger & T. Subhi Yamin (Eds.), *University Partnerships for Sustainable Development* (pp. 67-80). Emerald Publishing Ltd.

Hiser, K.K. & Lynch, M.K. (2021). Worry and Hope: What College Students Know, Think, Feel and Do about Climate Change. *Journal of Community Engagement and Scholarship*, 13(3), 96-107. https://digitalcommons.northgeorgia.edu/jces/vol13/iss3/7

Hoch, A., Stewart, D., Webb, K., & Wyandt-Hiebert, M.A. (May 2015). *Trauma-informed care on a college campus.* Presentation at the annual meeting of the American College Health Association, Orlando, FL.

House, L.A., Neal, C., & Kolb, J. (2020). Supporting the Mental Health Needs of First Generation College Students. *Journal of College Student Psychotherapy,* 34(2), 157-167. https://doi.org/10.1080/87568225.2019.1578940

Hungerford, H.R. & Volk. T.L. (1990). Changing Learner Behavior Through Environmental Education. *The Journal of Environmental Education,* 21(3), 8-21. https://doi.org/10.1080/00958964.1990.10753743

Kelly, R., Elsler, L., Polejack, A., van der Linden, S., Tonnesson, K., Schoedinger, S.E., Santoro, F., Pecl, G.T., Palmgren, M., Mariani, P., Glithero, D., Evans, K., Cvitanovic, C., Cooke, J., Bartram, J., & Wisz, M.S. (2021). Empowering young people with climate and ocean science: Five strategies for adults to consider. *SSRN.* https://dx.doi.org/10.2139/ssrn.3961791

Kim, S.H., Schneider, S.M., Kravitz, L., Mermier, C., & Burge, M.R. (2014). Mind-body Practices for Posttraumatic Stress Disorder. *Journal of Investigative Medicine,* 61(5), 827-834. https://doi.org/10.231%2FJIM.0b013e3182906862

Menschner, C. & Maul, A. (2016, April). Advancing Trauma-Informed Care Issue Brief: Key Ingredients for Successful Trauma-Informed Care Implementation. https://www.samhsa.gov/sites/default/files/programs_campaigns/childrens_mental_health/atc-whitepaper-040616.pdf

Morissette, S.B., Ryan-Gonzalez, C., Yufik, T., DeBeer, B.B., Kimbrel, N.A., Sorrells, A.M., Holleran-Steiker, L., Penk, W.E., Gulliver, S.B., & Meyer, E.C. (2021). The effects of posttraumatic stress disorder symptoms on educational functioning in student veterans. *Psychological Services,* 18(1), 124-133. https://psycnet.apa.org/doi/10.1037/ser0000356

National Child Traumatic Stress Network. (n.d.). *Introduction.* Retrieved April 7, 2022 from https://www.nctsn.org/trauma-informed-care/family-youth-provider-partnerships/introduction

Panu, P. (2020). Anxiety and the Ecological Crisis: An Analysis of Eco-Anxiety and Climate Anxiety. *Sustainability,* 12(19), 1-20. https://doi.org/10.3390/su12197836

Raja, S., Hasnain, M., Hoersch, M., Gove-Yin, S., Rajagopalan, C. & Kruthoff, M. (2015). Trauma Informed Care in Medicine: Current Knowledge and Future Research Directions. *Family and Community Health,* 38(3), 216-26.

Reyes, M., Carmen, B., Luminarias, M., Mangulabnan, S., & Ogunbode, C. A. (2021). An investigation into the relationship between climate change anxiety and mental health among Gen Z Filipinos. *Current Psychology,* 1–9. https://doi.org/10.1007/s12144-021-02099-3

Rogers, S.L., Jeffery, A.J., Pringle, J.K., Law, A.C., Nobajas, A., Szkornik, K., Turner, A.C., Moolna, A., & Hobson, L. (2021). The University Campus as a Learning Environment: the role of a Campus-based Living Lab in a Blended Teaching and Learning Environment. *Geoscience Communication*. Discuss. [preprint], https://doi.org/10.5194/gc-2021-32

Romeu, D. (2021). Is climate change a mental health crisis? *BJPsych Bulletin*, 45(4), 243–245. https://doi.org/10.1192/bjb.2021.30

Roysen, R. & Cruz, T.C. (2020). Educating for transitions: ecovillages as transdisciplinary sustainability "classrooms". *International Journal of Sustainability in Higher Education,* 21(5), 977-992. http://dx.doi.org/10.1108/IJSHE-01-2020-0009

Rukspollmuang, C., Mongkhonvanit, P., Phitthayanon, C., Silalai, N., & Nubsang, H. (2022). University as a Living Learning Lab for Sustainable Futures. *Journal of Sustainability Perspectives*, 2(1), 56-64. https://doi.org/10.14710/jsp.2022.15457

Silver, R.C., Holman, E.A. & Garfin, D.R. (2012) Coping with cascading collective traumas in the United States. *Nature Human Behavior*, 5, 4–6. https://doi.org/10.1038/s41562-020-00981-x

Stewart A. E. (2021). Psychometric Properties of the Climate Change Worry Scale. *International Journal of Environmental Research and Public Health,* 18(2), 494. https://doi.org/10.3390/ijerph18020494

Substance Abuse and Mental Health Services Administration. (2014a). *SAMHSA's Concept of Trauma and Guidance for a Trauma-Informed Approach*. HHS Publication No. (SMA) 14-4884. Rockville, MD. https://ncsacw.acf.hhs.gov/userfiles/files/SAMHSA_Trauma.pdf

Substance Abuse and Mental Health Services Administration. (2014b). *Trauma-Informed Care in Behavioral Health Services*. Treatment Improvement Protocol (TIP) Series 57. HHS Publication No. (SMA) 13-4801. Rockville, MD. https://www.ncbi.nlm.nih.gov/books/NBK207201/pdf/Bookshelf_NBK207201.pdf

Thomas, M.S., Crosby, S., & Vanderhaar, J. (2019, May 22). Trauma-Informed Practices in Schools Across Two Decades: An Interdisciplinary Review of Research. *Review of Research in Education*, 43(1), 422-452. https://doi.org/10.3102/0091732X18821123

Travers, A., Armour, C., Hansen, M., Cunningham, T., Lagdon, S., Hyland, P., Vallieres, F., McCarthy, A., & Walshe, C. (2020). *European Journal of Psychotraumatology*, 11, 1-10. https://doi.org/10.1080/20008198.2019.1708144

Wendt, S., Bastian, C., & Jones, M. (2021). Building Collaboration with Child Protection and Domestic and Family Violence Sectors: Trialling a Living Lab Approach. *British Journal of Social Work*, 51(2), 692-711. https://doi.org/10.1093/bjsw/bcaa206

Yang C., Chen A., & Chen Y. (2021) College students' stress and health in the COVID-19 pandemic: The role of academic workload, separation from school, and fears of contagion. *PLoS ONE*, 16(2). https://doi.org/10.1371/journal.pone.024667

The California State University Journal of Sustainability and Climate Change
2022, Volume 2

Development of Carbon Emission Assesment Tool Towards Promoting Sustainability in Cal State LA

Arezoo Khodayari (California State University, Los Angeles), Soumya Puvvada (California State University, Los Angeles), Jamaie M. Scott (California State University, Los Angeles), Chandra Mouli Rotta (California State University, Los Angeles), Geeth Sekhar Chadalawada (California State University, Los Angeles), Jivar Afshar (California State University, Los Angeles), Carlos D. Linares (California State University, Los Angeles), Brad Haydel (California State University, Los Angeles)

Abstract

The great demand for the burning of fossil fuels has greatly increased greenhouse gases (GHG) concentrations in the atmosphere. An increase in the atmospheric concentrations of greenhouse gases produces a positive climate forcing or warming effect [EPA, Climate Change Indicators]. Therefore, mitigation of GHG concentrations is important to prevent long-term impacts on the environment. On April 4, 2016, California State University, Los Angeles signed the most comprehensive of Second Nature's three Climate Leadership Commitments, the Climate Commitment. Following this commitment, California State University, Los Angeles, set the ambitious goal of operational carbon neutrality by the year 2040. To assist California State University, Los Angeles in moving effectively toward this goal, we developed an energy dashboard that can bring access, awareness, and education to campus about campus carbon footprint and promote energy-efficient behaviors. The developed energy dashboard is an interactive web application that works based on an energy model that is composed of various energy-consuming and GHG producing units such as Heating, Ventilation and Air Conditioning (HVAC), Heated Potable Water (HPW), Electricity, and Campus-Related Commutes. This energy dashboard enables individuals to analyze the campus's energy consumption and carbon footprint. Our research showed that campus-related commute was the first largest contributor to Cal State LA's carbon footprint in 2018 and accounted for 71.5% of carbon emissions. Electricity and heated potable water accounted for 20%, and 8.5% of the total campus carbon emissions, respectively.

Introduction

The Climate of Los Angeles is characterized by very mild, relatively rainy winters and hot summers [Climate and average monthly weather in Los Angeles, n.d.]. As our planet warms and shifts towards higher average temperatures during cold seasons, it can be anticipated that the demand for heating and usage of natural gas will reduce [Aroonruengsawat & Auffhammer, 2011]. However, the increase in extreme high-temperature events would likely lead to an increase usage of air conditioning systems [Aroonruengsawat & Auffhammer, 2011]. Both these scenarios would result in increased usage of electricity demand.

Universities are like small cities that are growing around the world. However, they do not have the grand operational scale and complex regulatory systems as cities do. Therefore, they have the advantage of mitigating greenhouse gasses emissions more easily within their boundaries. Since students, faculties and staff are not aware of the scale of the university's energy consumption or its resulting carbon footprint, they tend to waste energy [Yañez, Sinha, & Vásquez, 2019]. Therefore, to involve these individuals, self-monitoring intelligent dashboards have been created by many universities to support energy conservation [Martini, K. (n.d.)]. Such individuals' knowledge of energy consumption can prompt behavioral choices to reduce energy consumption whenever possible.

Universities around the world have been analyzing their energy use and carbon emissions and developing online energy dashboards to reduce their carbon emissions [de la Cruz-Lovera et al., 2017]. The Association for the Advancement of Sustainability in Higher Education (AASHE) reports that 70 of its member universities and colleges have used some forms of energy dashboards to communicate their energy consumption [AASHE, 2015]. For instance, UC Davis developed an Energy Dashboard in 2016, which analyzed energy data for buildings on the university campus to enable facilities management to improve energy efficiency with knowledge of and input into campus energy operations [Salmon, Morejohn, Pritoni, & Sanguinetti,2016]. Some universities conducted studies for longer periods. For example, the University of Almeria in Spain benchmarked the energy consumption data during a period of seven years alongside cross-sectional building information to establish a linear regression model to forecast future energy use [Chihib, Salmerón-Manzano, & Manzano-Agugliaro, 2020]. The Minnesota State University conducted a study to curb their emissions and control $CO2_e$ emissions by switching to LED lighting and upgrading the Central plant unit [Minnesota state University ESPC case study, 2021]. Their goal was to reduce the annual $CO2_e$ emissions by 4,462 metric tons and save about 400,000 dollars in energy costs over the 18-year contract. In addition, Miami University created a sustainability dashboard to focus on climate adaptation and community capacity building to deal with a changing climate. Their task force created a three-scope Climate Action Plan to mitigate the greenhouse gas emissions on campus. The three scopes are emissions directly produced on campus, emissions from purchased electricity, and emissions from commuting. They generated a Carbon footprint baseline to compare the future reductions and set a Climate Commitment for the University [Miami University, Sustainability Dashboard]. Analyses of this kind can help the university to identify its existing carbon emissions and to better assess the effectiveness of various energy-saving measures. This will result in more economical yet sustainable development on campus [Mohammadalizadehkorde, & Weaver, 2018].

To analyze the energy consumption on our campus, and to show the impact of various energy-saving measures on reducing campus energy consumption and carbon emissions we developed an interactive energy dashboard. The developed energy dashboard can be used to educate our campus community about campus' carbon footprint, to promote energy-efficient behaviors, and to reduce energy consumption on campus. Quantifying GHG contributions and reduction opportunities in various areas could further help to prioritize GHG reduction efforts. Although the university has adopted several low and zero carbon emissions measures such as utilizing solar panels, solar charging stations, and operating a hydrogen production and fueling station, it is still necessary to take further energy-saving measures to reduce the overall carbon emissions. In our analyses, energy consumption was divided into four major categories. We formulated energy consumption in each of these categories to make it possible to analyze the effect of operational variables on energy usage in each category. The analyses presented here are based on the data collected in 2018 and 2019, assuming similar operations in all the examined categories for both years. The sources of energy consumption in this research were broken down to heating, ventilation, air conditioning (HVAC), Electricity, Heated Potable Water or drinking water, and carbon emissions resulting from students, faculty/staff who commute to campus. It should be noted that due to the limited amount of hot water needed, potable water is heated in each building separately on our campus.

The interactive features of the developed energy dashboard would allow the user to assess the effectiveness of various energy-saving measures for reducing campus overall carbon emissions in the analyzed energy units. It should be noted that the practicality of these energy-saving measures would depend on both human, technical, and environmental factors [Agarwal, Weng, & Gupta, 2009]. The dashboard's visual interpretation of the data brings awareness and educates people regarding the carbon dioxide equivalent ($CO2_{e;}$ a standard unit for measuring carbon footprints) emissions from each module classified [Yun, et al. 2014].

Methodology

California State University, Los Angeles, has been in the heart of the city of Los Angeles since 1947 and is located 5 miles away from downtown Los Angeles. It is one of the Universities out of the 23 CSU Universities. It was ranked number one in the United States for the upward mobility of our students in 2017 [About the University, 2021]. In 2019, the university had about a total of 26,000 enrolled students and about 1,700 faculty [Workbook: Enrollment. (n.d.)., 2021]. The campus is approximately 175 acres and consists of more than 48 buildings that include parking spaces for the students, faculty/staff (Figure 1). In 2018 the total energy consumption in electricity was approximately 38,000,000 Kwh, and energy consumption in Heated Potable Water was around 16,000,000 Kwh.

The variables used in the energy formulation in each energy category were divided into three categories: Knobs, which are the variables that the user can adjust; Results, which are the variables that show the effects of changing the knobs; and Constants, which have constant values and are unchanged by alterations to the knobs. Typically, knobs are the input variables in the energy dashboard tool that can be adjusted to increase or reduce the intensity of the measuring object to determine the energy usage on campus (Arvind and Berry, 2015). Constants are the values that cannot be changed but

Figure 1
Campus Map of California State University, Los Angeles

can be seen by the Users for informational purposes. Finally, results are the output based on the input that users have incorporated into the dashboard. For example, a change in temperature (knob) would significantly change HVAC usage (result). This analysis will help us to know the contribution of various energy-consuming categories to energy consumption and carbon emissions on campus. The four main energy-consuming categories that were analyzed in this study are Transportation, Electricity, Heated Potable Water (HPW) and Heating, Ventilation, and Air Conditioning (HVAC).

Transportation

The transportation model analyzes the usage of energy based on the mode of transportation and the total number of gallons of fuel used by students, faculty/ staff commuting to California State University, Los Angeles. In this model, the data were broken down into two categories. Faculty /staff in one category and students in the other category. Further, these two categories were subdivided into drive-alone, rideshare, carpool, and public transportation. The data for students, faculty / staff commute habits to campus were collected through a campus-wide survey. The amount of energy used by commuting is complicated because it requires gathering a large amount of data. To simplify this complicated process and further assess and improve the data collected from the transportation survey, we used the number of daily and semester parking permits purchased in various parking lots on campus. In addition, to obtain information on how many students used public transportation, we used the number of purchased U-passes. U-pass provides unlimited public transportation to students who are enrolled as full-time students at the Cal State University of Los Angeles. The percentage of commuters using various modes of transportation and the total number of miles per commuter were collected from the survey. For this project, we collected the commute habits of students by conducting a campus-wide survey and commute habits of faculty/staff using the results of our campus, the South Coast Air Quality Management District (AQMD) Commuter Survey. The total enrollment of the students in 2019 was about 26,000, and the number of faculty was about 1700. We collected data from 1799 students and all faculty/staff. The results from the students' commuting survey were scaled up to the total number of enrolled students to calculate the total carbon emissions from the students' commute to campus. In the survey, various questions were asked, such as the number of days student/faculty/staff commute to campus, their zip code, commuter's mode of transportation for each trip during a week, and miles per gallon for the commuter's vehicle, etc. The transportation model multiplies the total number of miles commuted in each commute mode- obtained from the survey by the emission factor for that commute mode (kg of $CO2_e$ per mile). Then the result gets scaled up to the total population.

The Transportation model is defined by five variables. One variable is a Result (total energy used), three variables are Controls (distance, mpg, and unit conversion) and one variable is a Knob (number of trips). Together, these make it possible to model the energy used due to campus-related commutes.

The calculations can be modified in different ways depending on the selected transportation mode. For example, when students or faculty/staff carpool to campus, the resulted energy usage and carbon emissions is divided by the number of people who carpooled together. For public transportation like buses and Metrolink, the energy used can be calculated by a constant number (or factor) given by the U.S. Department of Transportation, which is recorded in Kg $CO2_e$ per mile per passenger. Then, it is multiplied by the total number of miles traveled by students using public transportation. The Kg $CO2_e$ per mile per passenger for the bus, is 0.224, and the Metrolink/Rail is 0.141 kg of $CO2_e$ per passenger in Southern California [Hodges, T, 2009]. We considered all the vehicles as passenger vehicles with 0.411 kg of $CO2_e$ per mile.

Electricity

Energy is frequently used directly in the form of electricity. This can be used in many ways in each building, from lights to computers to mini fridges. Due to the lack of monitoring of individual energy outlets, categorizing the energy usage depends on several approximations and assumptions. Several different variables could affect electricity usage. None of these variables are constants. All the variables are Knobs (number of lights, computers, walkways, personal devices, refrigerators, microwaves and projectors, and miscellaneous categories). One variable is a Result (total energy used). Using these variables together would make it possible to model the energy used as electricity. There is no granular-level data available for the electricity model, as the only data that we have is the metered reading of each building on the campus. But it cannot explain any breakdown of how much electricity was used by lights, microwaves, elevators, and other categories in a building. There was an extensive visual inspection done for three different buildings as a benchmark on campus to determine the number of lights, classrooms, offices, elevators, etc. Based on the class schedules obtained from the university scheduling offices, we did find the hours of operation of each classroom. Further, we made rough assumptions about the usage of other electricity-consuming devices. The buildings that were analyzed to see the energy consumption in different categories are King Hall, Salazar Hall, and Fine Arts. It is noted that this study only focused on electricity consumption in buildings and did not examine the carbon footprint of online instructions.

Heated Potable Water (HPW)

Potable water, or drinking water, is only heated for two circumstances: showers and sinks. Due to the limited amount of hot water needed, potable water is heated on an individual basis instead of at a central unit. On our campus, potable water is heated in each building separately. Each building has a water heater that runs on natural gas. Since we knew how much natural gas was used on campus to heat the water, we used that information to find the mass of water headed on campus. After calculating the mass of the water, we utilized it to calculate the energy used for heating the water as a function of other operational variables. In conclusion, the heated potable water model is defined by eight variables. Four variables are constant (efficiency of natural gas boilers, unit conversion factors, the specific energy of water, and incoming water temperature), two variables are knobs (the amount of water heated and the exiting water temperature), and two variables are results (the amount of energy and natural gas used).

The amount of monthly natural gas used by the entire university in the unit of therms was collected to obtain the energy consumption data for Heated Potable water. Certain assumptions were made to formulate the HPW model. We assumed that the efficiency of all the boilers based on their year and model would be around 82%. The data for the set temperature of the hot water and the temperature that the water goes into the heater were obtained from the facility personnel.

Heating, Ventilation, and Air Conditioning (HVAC)

The HVAC model depends on two factors: hours of operation and horsepower of the equipment in the air handling units. The horsepower of the equipment is the summation of the individual powers of each main piece of equipment in the HVAC system: The Return Fan, the Supply Fan, the Chilled Water (CHW) Pump, and the Compressors. These are the components present in each building. The Central Plant is where the cooling takes place, and its energy consumption is obtained from the meter reading. The hours of operation are based on the external temperature Monday through Saturday. The system is off on Sunday. The HVAC system remains turned off when the external temperature is lower than the thermostat's set temperature. The HVAC system turns on when the external temperature is higher than the thermostat's set temperature. The external temperature is higher during summer, resulting in higher hours of operation.

Similarly, the external temperature is lower during winter, resulting in lower hours of operation. The data for the HVAC model were obtained from three different areas. First, the power data were collected from the horsepower of the individual equipment in the air handling unit in each building. For example, the Return Fan, Supply Fan, CHW Pump, and Compressors each have associated horsepower, which varies in each building. Second, the hours of operation were determined based on the external temperature. Finally, the energy spent in the Central Plant to cool down the water to provide chilled water was determined from the meter reading. Together, these values enabled us to formulate our HVAC model.

The Heating Ventilation and Air Conditioning system (HVAC) energy consumption is not only affected by the operational hours but also by the external temperature. Therefore, HVAC energy usage is much higher in warmer months than in colder months. The HVAC model was formulated based on the equipment in the HVAC system and the number of daily operating hours. The number of hours per day was based on the number of hours that the external temperature was above the thermostat set temperature (i. e. 73 Degrees Fahrenheit), not the number of hours school was in session. Since the HVAC in all universities is set to a specific temperature where there is an automatic turn-on without human interaction, the main factor contributing to this model was the outside temperature.

Result and Discussion

Table I demonstrates the results of our analysis for the transportation model. Our analyses show that 95% of energy consumption/carbon emissions in the campus-related commutes comes from students' commute and 5% from faculty/staff commute. While the data used by the model is self-reported, it can still be seen as highly accurate. The commute data is reported by students and faculty/staff by completing a survey that is sent annually. It should be noted that volume of data obtained each year has increased due to additional surveys submitted by students and staff/faculty which resulted in an accurate analysis. For example, the annual average carbon emissions due to commuting to campus in 2019 were calculated to be approximately 58865 Ton of CO_2e for students and 2965 Ton of CO_2e for faculty/staff.

This gives an opportunity to show a general concept of how much energy is used by students, as well as shows possibilities for reduction (carpool, bus, etc.).

Figure 2 illustrates the breakdown of electricity usage in one of the analyzed buildings (i. e., King Hall building) as an example of the relative difference in the magnitude of electricity usage in various subcategories of the electricity model. The electrical energy consumed in king hall per month is approximately 172,794 Kwh. The results were compared with the meter reading, and the error was below 5%. The analyses for other buildings are available on our energy dashboard website.

This electricity model is highly accurate but will require the most maintenance to remain so. As of now, the model is based on a current walk through of campus showing the items in use, and the energy used by those items. Due to the rapid growth of technology, many of these items are changed regularly as better items become available (new computers, projector instead of blackboard, etc.). For this model to be maintained accurately, this will need to be updated on a regular basis.

Figure 3 shows the percentage of monthly energy consumed to heat the potable water related to the total annual energy consumed in the HPW category (1.61×107 Kwh) in 2018. The weather temperature in February, March, and April is cooler, resulting in higher demand for the heated water. In addition, school is in session during these months, which means there are more students on campus. As you can see in Figure 4. the energy consumption is higher in February, March, and April. The energy combustion decreases in the hotter months from May to September. However, it was observed that the energy demand increased slowly after September but

Table 1
Annual Average CO2e Emissions Per Commute Mode

Commute Mode	$CO_{2,e}$ Emissions (Ton $CO_{2,e}$)
Drive Alone/Drop Off	53812.85
Rideshare (Uber/Lyft)	2565.82
Carpool	2483.63
Bus	2.27
Metrolink/Rail	0.99

Figure 2
Electricity Consumption for King Hall Building (KWh)

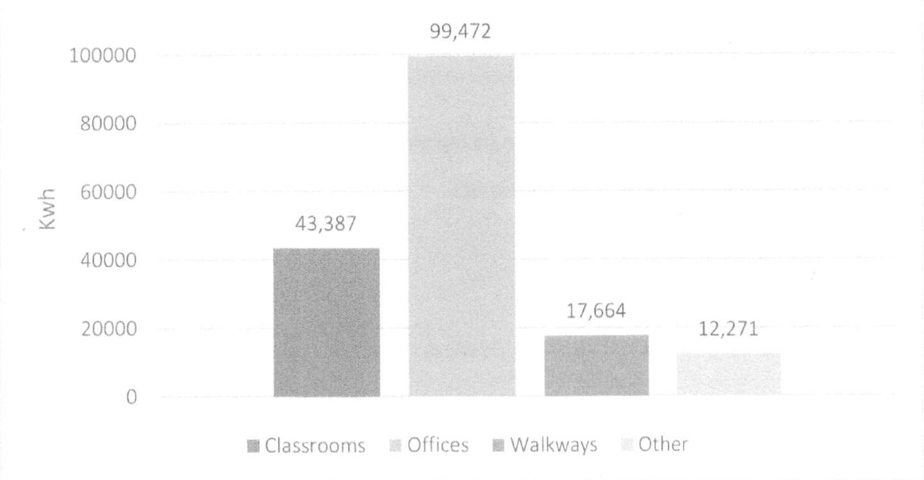

was less in December and January. The lower level of energy combustion in December and January is due to the winter break during that time. As was expected, there is very little energy used by heating potable water, so this accounts for a very small percentage of the campus energy usage overall.

The HVAC results are directly affected by the California's Mediterranean-like climate, where the summers are dry and hot, and winters are humid and freezing temperatures are rare. These weather conditions increase the use of HVAC, and the contribution of the HVAC is more than HPW [("Monthly weather forecast and climate Los Angeles, CA," 2021)]. Most of the university buildings are poorly ventilated with windows which results in no airflow from outside to inside through windows as all of them are closed. The airflow to keep the building cold during summers is through HVAC as there is no other way that the facility is held at that temperature. Based on our analyses the HVAC model accounts for a large percentage of the energy used on campus, which is the result of Los Angeles climate. Our results indicated that the energy used for each building's HVAC was between 20%-50% of

Figure 3

Energy Consumption in the Heated Potable Water Category in 2018 in Kwh

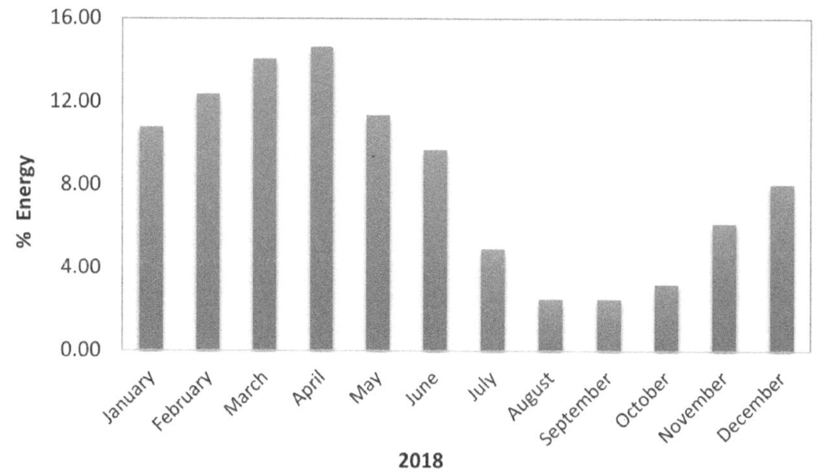

Figure 4

HVAC Annual Average Energy Consumption per Building in 2018

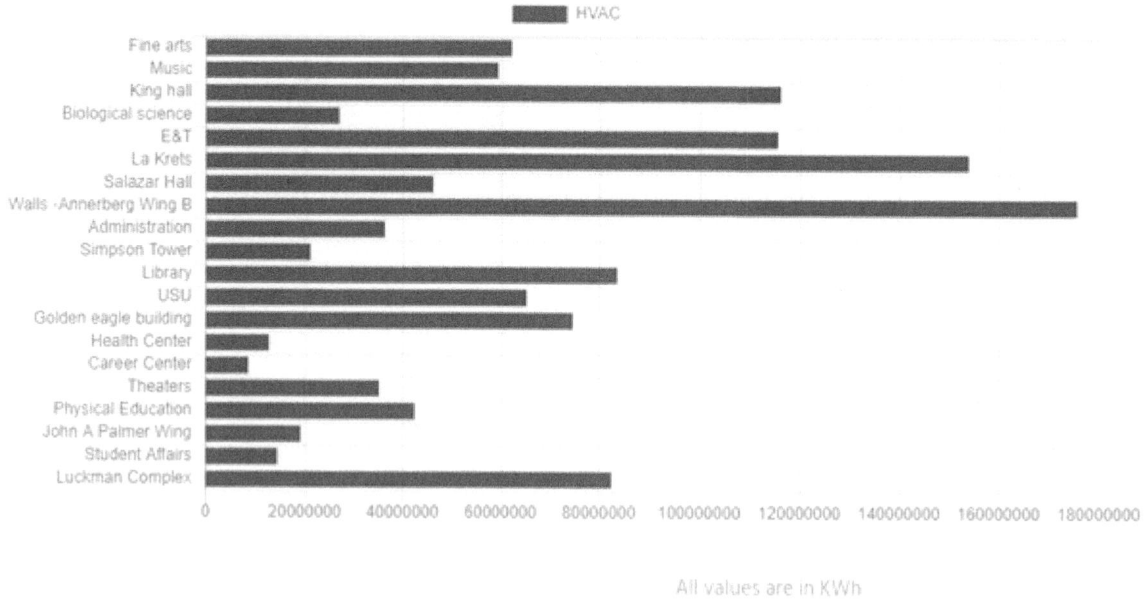

the total energy used per building. Figure 4 shows the HVAC energy consumption per building.

Overall, the HVAC system would represent a great opportunity for energy reduction as its contribution to energy consumption is relatively high and there are various ways to reduce its energy consumption (e.g., adjusting the schedule for course offerings, lowering the HVAC set temperature, etc.).

Finally, the carbon footprint analyses of all the analyzed energy-consuming sectors revealed that campus-related commute, total electricity usage (electricity consumption for lighting and HVAC), and heated potable water account for about 71%, 20%, and 9% of the total campus carbon emissions, respectively. It is noted that uncertainties were not considered in this study. Figure 5 shows the CO_2 emissions equivalent (Gg $CO2_e$) from different Energy-Consuming Categories in 2018. As illustrated in Figure 5 campus related commute is the largest energy consuming activity and contributor to carbon emissions. Generally, commuting-related activities are a major component of many institution's carbon emissions, and our analyses confirms this as well. This result highlights the importance of reducing campus-related carbon emissions by providing hybrid instruction (i.e., a combination of in-class and online learning) and promoting public transportation and carpooling by offering certain incentives.

Conclusion

After analyzing all the four model results for 2018, we concluded that students, faculty/ staff's commute to campus has the most substantial contribution to campus-related carbon emissions. Campus-related carbon emissions can be reduced by taking various measures such as changing lightbulbs with more energy-efficient options, increasing the set temperature in the HVAC system by few degrees, providing online/hybrid instruction and promoting public transportation and carpooling by offering certain incentives. California State University, Los Angeles campus has been working actively in all these areas to reduce its carbon emissions and promote sustainable behaviors. With the use of our interactive energy dashboard, the campus community should be able to better track and compare the energy consumption of different campus-related activities and further assess the impact of taking certain energy-saving and carbon-reducing measures on the reduction of the campus's overall carbon footprint. Also, the campus carbon footprint can be further reduced by utilizing renewable energy and purchasing carbon credits. The campus has already started utilizing renewable energy and low-carbon energy by using solar panels, operating a hydrogen fueling station on campus, and purchasing carbon credits. By taking various energy-saving measures, utilizing more renewable and low-carbon energy, we will reduce our carbon footprint and move toward a more sustainable campus.

Acknowledgements

This work has been supported by Gunjit S. Sikand Faculty Endowment to Cal State LA College of Engineering, Computer Science and Technology (ECST) for research in urban sustainability. Authors would like to thank the Cal

Figure 5
CO2 Emissions (Gg CO2e) from Different Energy-Consuming Categories in 2018

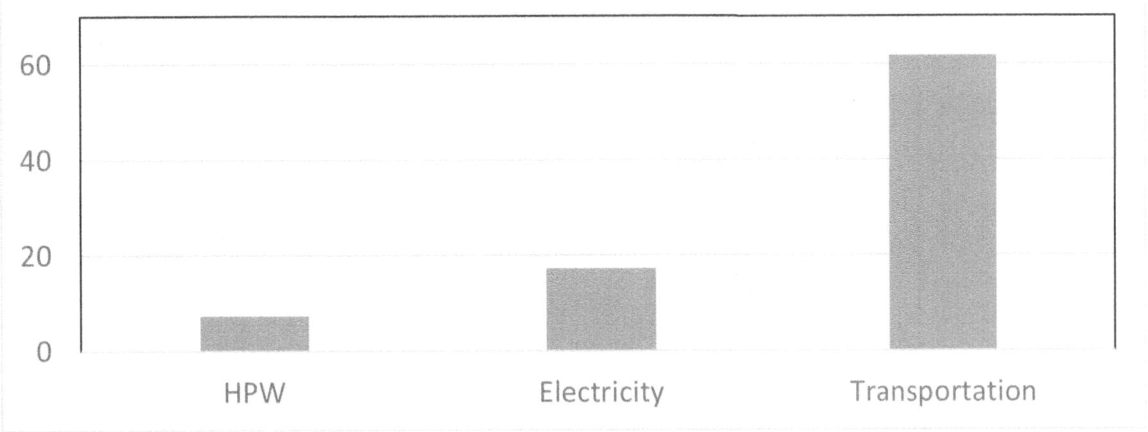

State LA Energy and Sustainability Office for their assistance in data collection.

References

About the University. (2021, March 01). Retrieved March 29, 2021, from https://www.calstatela.edu/about

Agarwal, Y., Weng, T. & Gupta, R. (2009). The energy dashboard: Improving the visibility of energy consumption at a campus-wide scale. *Proceedings of the 1st ACM Workshop Embedded Sensing Systems for Energy-Efficiency in Buildings*, pp. 55-60. https://doi.org/10.1145/1810279.1810292

Anin Aroonruengsawat & Maximilian Auffhammer, 2011. "Impacts of Climate Change on Residential Electricity Consumption: Evidence from Billing Data," NBER Chapters, in: *The Economics of Climate Change: Adaptations Past and Present,* pages 311-342, National Bureau of Economic Research, Inc.

Arvind Jense and Berry Eggen. (2015). *Awakening the Synthesizer knob: Gestural Perspectives.* Retrieved March 31,2021, from https://www.researchgate.net/publication/283636364_Awakening_the_Synthesizer_Knob_Gestural_Perspectives

Chihib, M., Salmerón-Manzano, E., & Manzano-Agugliaro F. (2020). Benchmarking Energy Use at University of Almeria (Spain). *Sustainability*, 12(4), 1336. https://doi.org/10.3390/su12041336

Climate and average monthly weather in Los Angeles (California), United States of America. Retrieved 23 April 2022, from https://weather-and-climate.com/average-monthly-Rainfall-Temperature-Sunshine,Los-Angeles,United-States-of-America

Cruz-Lovera, C.D., Perea-Moreno, A., Cruz-Fernández, J.D., Álvarez-Bermejo, J.A., & Manzano-Agugliaro, F. (2017). Worldwide Research on Energy Efficiency and Sustainability in Public Buildings. *Sustainability*, 9, 1294. https://doi.org/10.3390/su9081294

Environmental Protection Agency. *Climate Change Indicators: Greenhouse Gases.* EPA. Retrieved April 17, 2022, from https://www.epa.gov/climate-indicators/greenhouse-gases

Hodges, T. (2009). Public Transportation's Role in Responding to Climate Change. FTA. Washington, D.C. http://www.fta.dot.gov/documents/PublicTransportationsRolInRespondingToClimateChange.pdf

Khodayari, Arezoo et. al. (2020). *Energy Dashboard.* https://cysun.org/espc-researchlab/EnergyDashboard/

Martini, K. (n.d.). *The effects of Energy dashboards and Competition programming on CYCLIC electricity consumption on a college campus.* Retrieved March 30, 2021, from https://scholarworks.wmich.edu/dissertations/3545/

Minnesota state University ESPC case study. (2021, March 03). Retrieved March 31, 2021, from https://www.ameresco.com/portfolio-item/minnesota-state-university-mankato/

Mohammadalizadehkorde, M. & Weaver, R. (2018). Universities as Models of Sustainable Energy-Consuming Communities? Review of Selected Literature. *Sustainability* 2018, 10, 3250. https://www.mdpi.com/2071-1050/10/9/3250

Monthly weather forecast and climate Los Angeles, CA. (2022). Retrieved March 31, 2021, from https://www.weather-us.com/en/california-usa/los-angeles-climate

Salmon, K., Morejohn, J., Pritoni, M. & Sanguinetti, A. (2016). The Iterative Design of a University Energy Dashboard. In: *ACEEE Summer Study on Energy Efficiency in Buildings.* http://aceee.org/files/proceedings/2016/data/papers/8_466.pdf

Sustainability Dashboard. Dashboard | Snapshot | Sustainability - Miami University. (n.d.). Retrieved April 18, 2022, from https://www.miamioh.edu/about-miami/sustainability/dashboard/index.html

The Association for the Advancement of Sustainability in Higher Education (AASHE) (2015). Campus Building Energy Dashboards. http://www.aashe.org/resources/campus-building-energy-dashboards

Workbook: Enrollment. (n.d.). Retrieved March 29, 2021, from https://latabpubsrc.calstatela.edu/t/LAIR-PUBLIC/views/Enrollment/AQuickGlance?%3Aembed_code_version=3&%3Aembed=y&%3AloadOrderID=0&%3Adisplay_count=n&%3Adisplay_spinner=no&%3AshowVizHome=n&%3Aorigin=viz_share_link

Yañez, P., Sinha, A., & Vásquez, M. (2019, December 24). *Carbon footprint estimation in a university campus: Evaluation and insights.* Retrieved March 30, 2021, from https://www.mdpi.com/2071-1050/12/1/181

Yun R. et al. (2014) The Design and Evaluation of Intelligent Energy Dashboard for Sustainability in the Workplace. In: Marcus A. (eds) *Design, User Experience, and Usability. User Experience Design for Everyday Life Applications and Services.* DUXU 2014. Lecture Notes in Computer Science, vol 8519. Springer, Cham. https://doi.org/10.1007/978-3-319-07635-5_58

The California State University Journal of Sustainability and Climate Change
2022, Volume 2

Understanding Consumers' Motivations and Product Preferences for Deconsumption

Anubha Mishra (Rider University), Norm Borin (California Polytechnic State University - San Luis Obispo), Joan Lindsey-Mullikin (California Polytechnic State University - San Luis Obispo), Ram Krishnan (California Polytechnic State University - San Luis Obispo)

Abstract

The modern era features dramatic changes in consumer attitudes and behaviors related to consumption; this article investigates the motivations and product preferences of consumers who alter their consumption habits to obtain products that cause less damage or use fewer resources. The authors offer an empirical definition of such deconsumption and compare it with other terms used in prior literature to refer to resource-restricted purchases; they also propose consumer-oriented measures of deconsumption motivations and product preferences. The results reveal ten motivation factors that help explain why people choose to deconsume. Regression analysis further shows that product choice varies across these motives. Thus, businesses, public policy makers, and non-profit organizations can modify their communication and product strategies, for both green and conventional products and services, to appeal to different subgroups of consumers that deconsume for distinct reasons.

Introduction

Discussions related to climate change, resource depletion, and global pollution continue to increase in frequency and intensity. If we meet the predicted world population increase for the next two decades, we will require the natural resources equivalent to three Earths to sustain our current lifestyles (UN, 2022). Consumer household purchases play a significant role in resource depletion, increased waste, and rising greenhouse gas emissions (Long et al., 2022; Marini et al., 2022; Piligrimiene et al., 2020; Sun et al. 2022). In response, the modern era is marked by changes in consumer attitudes and behaviors related to consumption, in that consumers are more mindful of the impact their consumption choices have on the environment and society, and they actively seek products that are less damaging, use fewer resources, or are produced in fair labor conditions. Consumers who embrace such consumption standards take various labels, such as responsible (ORCI, 2012), green (Janssonet al., 2010), environmentally conscious (Zabkar & Hosta, 2013), sustainably conscious (Hartman Report on Sustainability, 2007), and socially responsible (Durif et al., 2011).

Another group of consumers seeks to reduce consumption levels in general, often referred to as anti-consumers (Black & Cherrier, 2010), voluntary simplifiers (Etzioni, 1998), brand avoiders (Iyer & Muncy, 2009), or mindful consumers (Sheth et al., 2011). Extant research into the attitudes and behaviors of such consumers often centers on the reasons for their anti-consumption (Black & Cherrier, 2010; Iwata 1997, 1999) to seek a complete sense of consumption. Anti-consumers seek simpler lives, less material consumption, less work, more time with their family, and reduced impacts on the environment (Johnston & Burton, 2003). In this stream, other studies propose segmentation standards that reflect these consumer attitudes and behaviors (Etzioni, 1998; Iyer & Muncy, 2009), using the reduced consumption effort as the focal topic. But these consumers still consume, so an interesting further area of analysis relates to the products they choose to purchase. Researchers predict that such consumers prioritize the functional aspect of products (Craig-Lees & Hill, 2002) or look for products with meaning (Black & Cherrier, 2010), and some studies indicate their overlap with green consumers and identify a specific segment of pro-social consumers (i.e., deconsumers) (Lee et al., 2009).

Table 1 summarizes a variety of perspectives and some of the commonly used descriptive terms related to reduced consumption. However, prior literature provides descriptors of reduced consumption with unique motivations, as well as distinct behaviors that reflect these motivations. Most quantitative research leaves the definition of the behaviors related to reduced consumption up to the individual consumer, leaning heavily on behavioral measures associated with environmental concerns, such as recycling, biking, and buying organic foods. We seek to advance this research domain by proposing a precise definition of deconsumption, based on both prior research and consumer input. The current study relies on the term "deconsumers" to emphasize that the consumers being studied still consume but also avoid any adversarial connotations (e.g., as arise with terms like anti-consumption).

In contrast with some prior research in table 1, we also focus on deconsumption motivations and purchases, as opposed to lifestyles and other tangential activities such as, decreasing the number of hours worked, changing job locations or helping disadvantaged groups. Our focus on consumption provides concrete examples of the types of product categories that can help consumers decrease their overall consumption. Further, to expand prior literature, we address the product preferences and consumptive behaviors related to different deconsumption motivations. This investigation of the link between consumers' motivations and product and behavioral strategies helps us answer more niche questions such as, what types of products do people who deconsume for simplification reasons select? Do they differ from the products preferred by those who deconsume for

Table 1
Commonly Used Descriptors Related to Reducing Consumption

Consumer Description	Definition of Segment and Connection with the Current Research
Voluntary Simplifier	Some research focuses on the consumption aspect: "the choice out of free will rather than by being coerced by poverty, government austerity programs, or being imprisoned to limit expenditures on consumer goods and services, and to cultivate non-materialistic sources of satisfaction and meaning" (Etzioni, 1998, p. 620). Other research examines this group more broadly, investigating lifestyle aspects such as turning down promotions, working part time, reading more, or watching less television (Iwata, 1997).
Anti-Consumption	This label describes people who are against consumption, either selectively or across the board: "literally means against consumption" (Lee et al., 2009, p. 145) "A practice of rejection, reduction and reuse" (Black & Cherrier, 2010, p. 438) Related research focuses on what these consumers will not consume, rather than what they might consume.
Mindful Consumption	This recently introduced term implies a "Consumer Mindset of caring for self, for community, and for nature that translates behaviorally into tempering the self-defeating excesses associated with acquisitive, repetitive and aspirational consumption" (Sheth et al., 2011, p. 21). These authors introduce a framework and recommendations to companies to use the consumer's mindset in developing their strategies. At this point, the framework has not been tested.
Socially Responsible Consumption	"A person basing his or her acquisition, usage, and disposition of products on a desire to minimize or eliminate any harmful effects and maximize the long-run beneficial impact on society" (Mohr et al., 2001, p. 47) Early studies focused primarily on the environmental impacts of consumption. More recently, social dimensions have been added (Webb, Mohr, and Harris 2008). Studies have not focused on connecting motivations to subsequent product choices that coincide with these motivations.

environmental reasons? Etc. Overall, this study aims to address the lack of comprehensive exploration of research on reduced consumption. Three main contributions of the research are:

1. Introduce an empirically generated definition of deconsumption based on a survey designed to gather qualitative consumer input on its meaning.
2. Identify distinct reasons that people deconsume, on the basis of prior literature and consumers' inputs, which guides the development of motivation statements that can evaluate why consumers choose to deconsume.
3. Connect the reasons people deconsume with different product/behavioral strategies that they use to satisfy their deconsumptive needs.

We start with a review of relevant research that investigates consumers' interests in reducing consumption.

Deconsumers: Attitudes, Behaviors, and Segment Descriptions

Traditionally, consumption offers a means to express individual identities, values, and beliefs (Grinstein & Nisan, 2009). Simplistic consumption is not just a coping behavior but has become a preference for many consumers; even affluent consumers note their dissatisfaction with excessive consumption and seek to lead a less wasteful life (Flatters & Willmott, 2009). Despite this notable appeal of deconsumption for consumers, the perceived costs and lack of immediate evidence of environmental benefits make it difficult for consumers to change their behaviors completely, especially if the changes go beyond using green products or avoiding some products altogether. Therefore, we seek to explore the various motivations that lead consumers to deconsume.

Sustainable consumption research often attempts to explain people's attitudes or behaviors with a variety of terms; a common description referred to voluntary simplifiers (VS). For example, Iwata's (1997) VS scale suggests that these consumers only purchase necessary items, want to simplify their lives, focus on personal fulfillment rather than material purchases, and are environmentally concerned. As noted previously, other terms also have been used to describe deconsumers, such as socially responsible, ethical, or green, and some behaviors displayed by these groups are deconsumptive. However, they also might simply redirect their consumption from one product to another (e.g., electric car as an alternative to an internal combustion engine car; Sheth et al., 2011). In this sense, anti-consumption investigations may be more relevant for our study, because they pertain to people who fight against consumption (Lee et al., 2009). Iyer and Muncy (2009) classify four anticonsumer groups according to whether they avoid all types of products or only selected ones and whether they avoid consumption for personal or societal reasons. Their brief, eight-item survey offers some initial confirmation that anti-consumptive groups vary by their reasons for anti-consuming.

Friends, family, and other external group associations also can influence individual deconsumptive motivations and behaviors. Social identity, which traditionally stems from intergroup relations, has a tremendous influence on attitudes and subsequent behaviors related to climate change and the environment. When people seek a particular social identity associated with a group whose norms are environmentally responsible, they are more likely to engage in pro-environmental decisions and behaviors too (Fieldling & Hornsey, 2016). Similarly, collective efficacy manipulations increase proenvironmental intentions at collective and individual levels, such that a greater sense of efficacy grants people more motivation to join in collective pro-environmental behaviors and display similar individuallevel behaviors (Jugert et al., 2016).

Reflecting our understanding of extant research related to consumer interests in reducing consumption, we first conceptualize deconsumption, then present the methodology we used to extract the factor structure for deconsumption motives and strategies. The next section contains an analysis of the relationships of deconsumption motivations with strategies. Finally, we offer practical insights and conclude with a discussion of future research directions.

Methods

Conceptualizing Deconsumption

As discussed, the variety of terminologies used to investigate different aspects of deconsumption implies that extant research has not reached a consensus regarding the meaning of deconsumption. Therefore, developing a deeper understanding of this term is an important contribution of the current study. With a national, web-based survey, we asked 90 respondents to define deconsumption and provide examples of deconsumptive behavior. An analysis of these data suggests that people believe that deconsumption stems from many different motives, such as simple living, money management, environmental concerns, or health concerns. Table 2 includes some example excerpts from these respondents. Accordingly, we define a deconsumer as follows: "Consuming products or services uses energy, food, or materials; a deconsumer is a person who modifies his or her consumption to use less energy, food, or materials over time."

Developing the Measurement Instrument

Using the qualitative comments and existing research, we developed an item pool to investigate consumers' motivations for deconsumption. Reflecting a combination of insights from the comments and existing scales (e.g., Iwata, 1997, 1999), such as, the Green Consumer Values (Haws et al., 2013), Socially Responsible Consumption Behavior (Antil, 1984), Voluntary Simplicity (Cowles & Crosby, 1986), and Simplifiers and Global Impact Consumers (Iyer & Muncy, 2008), the item pool explores consumers' various motivations for deconsumption.

Again, turning to extant literature, survey comments, as well as feedback from academicians, we developed measures for potential product/behavioral deconsumption strategies. Some product options reflect functional aspects (Craig-Lees & Hill, 2002), whereas others refer to more emotional meanings or peer-to-peer connections that might be achieved from consuming a product (Black & Cherrier, 2010).

Furthermore, some describe a green feature as inherent to the product (Lee et al., 2009).

The survey included 50 items, on a five-point agreement scale, to measure motives for deconsumption. To measure product/behavioral deconsumption strategies, we asked respondents about the frequency with which they deconsumed, by using various products or behaviors. These 22 items were also measured on a five-point scale (1 = "never"; 5 = "always"). Finally, demographic variables, such as gender, age, income, ethnicity, and education, served as controls.

Procedure and Participants

We used a web survey among Amazon Mechanical Turk (MTurk) participants who were United States residents and 18 years or older. Of the total 622 responses, we deleted 122, due to excessive case-level missing data (>50%), straight-lining responses (e.g., 1 or 5 responses for the entire survey), or respondents who exited the survey after the introduction screen. The total useable sample size is 500.

In the sample collected, gender was almost equally represented (49% women). Nineteen percent of the sample was between 18 and 34 years of age, 20% were 25–34 years, 26% were 35–44 years, and 35% were 45–65 years or older. Approximately 25% of the respondents had annual household incomes above $70,000, and the sample was predominantly

Table 2
Meaning of Deconsumption

What Does Deconsumption Mean to You?	Quotes From Respondents
Simple living	Living modestly. Not buying things I don't need. Less cluttered home which leads to a sense of freedom. Reducing the amount of unecessary things or activities you have in your life. In a way, it is simplifying your life of all extraneous things and activities that do not enhance your life at all.
Environmental concern	An individual that works toward shrinking his or her carbon footprint. Buying food that is environmentally friendly.
Monetary concern	Lowered their spending and consumption to save money and generate less waste.
Health concern	Not using/purchasing items that could be harmful to yourself or other people. Using organic products to have healthier children.
Community buying	Pooling resources and buying something as a small group, and sharing use of that something. Example: resident-owned housing cooperative. Another example: three or four neighbors with tiny, tiny lawns share a lawn mower.
Philosophical	Using only what you NEED and not all you want. Positive influence on the people around you by increasing the positive impacts of consuming.
Other	Buying used products, selling items used instead of throwing them away, closely watchin resource use and thinking of ways to reduce it.

Caucasian (77%). Finally, 99% of the sample had at least completed high school, 46% were college graduates, and 15% had received a graduate degree.

Statistical Analyses

Before testing the relationship between consumers' motivations and strategies for deconsumption, we conducted an exploratory factor analysis (EFA) to identify the latent constructs underlying the measured variables followed by a confirmatory factor analyses (CFA). The following dis-cussion details the procedure applied to assess motivations for deconsumption; we used a similar procedure to analyze deconsumption strategies.

With SPSS, we subjected the motivation scale of deconsumption to an EFA with principal component extraction and Varimax rotation. Factors with eigenvalues greater than 1 were retained. Indicators with factor loadings larger than .5 or cross-loadings of .3 or less remained in the further analysis (Costello & Osborne, 2005). Through this process, we dropped 18 items.

Next, we ran a CFA to analyze the final set of items corresponding to each factor in the measurement model. For the most part, the CFA results in table 3 support the factor structure identified by the EFA. For the measurement model, the key statistics $\chi 2$ = 546.21 (df = 389; p ≤ .00), confirmatory fit index (CFI) = .97, incremental fit index (IFI) = .97, and root mean square error of approximation (RMSEA) = .039, indicate good model fit and unidimensionality (Anderson & Gerbing, 1988).

We tested for convergent and discriminant validity, according to the average variance extracted (AVE) and composite reliability. The reliabilities range from .64 to .91 (Fornell & Larcker, 1981), and the AVE for each construct is greater than 40%. Given the exploratory nature of this research, we considered the major factors according to the need for plausibility to identify sufficient common factors (Fabrigar et al. 1999; Fava & Velicer, 1992), thereby retaining factors with AVE of 40% and above. When we analyze deconsumption strategies, the items produce three factors. The CFA results for overall model fit are as follows: $\chi 2$ = 1191.29 (df = 91; p ≤ .00), CFI = .95, IFI = .94, and RMSEA = .054. Tables 3 and 4 show the item loadings and CFA statistics for motivations and strategies, respectively.

Results

Deconsumption Motivations

The analysis reveals ten distinct deconsumption motives. The first factor describes saving resources and disliking waste. Consumers motivated by this factor are worried about others as much as themselves, similar to Iyer and Muncy's (2009) global impact consumers, rather than simplifiers (who tend to be more internally focused). This deconsumption motivation factor is most strongly aligned with the characteristics of segments described in previous research, such as true-blue greens or greenback greens (Roper, 2002) or the lohas or resource conservers (Ottman, 2017). We refer to this factor as ***resource concerns***. Another factor signals motivations that closely match general descriptors of VS in prior literature, so we name it ***simplifying***. Simplifying-based reasons for deconsuming are more internal and focused on downscaling, reducing clutter, or not being materialistic. These consumers tend to both buy and spend less.

Two other factors pertain to motivations related to a company's activities or personal health. The motivations for the factor we call *corporate objections* are similar to those assigned to Iyer and Muncy's (2009) market activists, including disapproval of products from companies that pollute or whose behaviors fail to align with the individual consumer's interest in consuming less. Similarly, a set of health-related motivations for deconsumers, termed ***health concerns***, appears both internally and externally focused, derived largely from a desire for clean living and avoiding harmful products. Relative to extant research, these motivations probably match most closely with classic strong environmentalists, such as true-blue greens (Roper, 2002).

Three other factors entail bonding with the community, sharing resources, or monetary constraints. First, items measuring a strong sense of community and desire to associate with like- minded others load on a single factor, which we term ***community bonding***. This interesting factor refers to deconsumption behavior, as influenced by other deconsumers. We thus infer that such motives are social in nature (i.e., consume products to achieve a sense of belonging). Second, the ***sharing*** factor comprises items that measure the motivation to share products and beliefs in the benefits of deconsumption by buying less. Third, ***monetary constraints*** indicate affordability as the main reason to deconsume. We believe it is important to include this factor in our analysis, to test the general perception that some people choose to reduce their consumption for financial reasons, rather than environmental concerns or other deconsumptive motivations.

Finally, in contrast with simplifying, frugality motives reflect desires to reduce the stress associated with spending too much. Two other factors, religious concerns and political concerns, indicate the influence of people's beliefs on their deconsumption behavior. Both of these factors comprise

Table 3
Confirmatory Factor Analysis of Deconsumption Motivation

I deconsume because I believe...	FL	AVE	CR
Resource Concerns		50.44%	.86
if we could use a little less there would be more left for future generations.	.79		
we should limit our use of products made from scarce sources.	.64		
every person should reduce their buying of products so sources can last longer.	.71		
natural resources must be preserved even if we must do without some products.	.65		
we must all do our part to consume less.	.74		
if we all consume less, the world would be a better place.	.72		
Simplifying		45.13%	.71
in living a simple life by not buying articles which are not necessary.	.67		
I put less emphasis on material things than most people know.	.73		
not buying a lot of things helps me declutter my life.	.61		
Corportate Objections		64.91%	.88
in not buying products from companies that engage in activities that I do not agree with.	.79		
in avoiding products that come from companies whose products do not stand for something I believe in.	.76		
in not buying products from companies guilty of polluting the environment even though it might be inconvenient.	.84		
in avoiding products from a company that I know may be harming the environment.	.83		
Health Concerns		45.34%	.71
in avoiding household chemicals that are not environmentally friendly.	.74		
in avoiding fruits and vegetables grown with pesticides or chemicals.	.58		
in not buying products that are harmful to those around me.	.69		
Community Bonding			
in purchasing items that I think other deconsumers will approve of.	.79		
I achieve a sense of belonging by consuming like other deconsumers do.	.83		
my consumption behavior is guided by my desire to associate with deconsumers.	.79		
Sharing		60.85%	.76
in sharing items, such as bikes, cars, hedgers with others.	.79		
it makes sense for multiple people to share the benefit of an item without everyone buying one.	.77		
Monetary Constraints		46.93%	.64
I can't afford to buy lots of things.	.69		
my financial constrains reduce my ability to buy the things I want.	.68		
Religious Concerns		76.88%	.91
my religious beliefs are what lies behind my approach to deconsumption behavior.	.90		
that excessive buying is against my religious beliefs.	.87		
I try hard to carry my religion over into all other dealings in life such as deconsumption behavior.	.86		
Frugality		77.93%	.88
it reduces my financial stress.	.81		
it saves me money.	.95		

I deconsume because I believe...	FL	AVE	CR
Political Concerns		64.26%	.83
my political beliefs are what lies behind my approach to deconsumption behavior.	.82		
that excessive buying is against my political beliefs	.77		
I try hard to carry my political beliefs over into all other dealings in life such as deconsumption behavior.	.78		

FL = factory loading; AVE = average variance extracted; CR = compose reliability.

Table 4
Confirmatory Factor Analysis of Deconsumption Strategies

How often do you deconsume by...	FL	AVE	CR
Green Product Search		47.71%	.88
switching to products that-			
produce less carbon footprint	.80		
create less waste	.77		
are less harmful to the environment even if that means buying the same number of products	.74		
use less material	.72		
use less water	.67		
use less energy	.61		
buying a product that-			
has been significantly recycled from other products	.65		
can be recycled into something else (e.g., glass, paper, or some types of plastics can be made into other products)	.51		
Temporal Product Search		42.81%	.69
reusing discarded products of others (e.g., garage sale, Goodwill)	.71		
sharing or renting out something you own (e.g., Airbnb; sharing tools with others)	.63		
making my own products	.62		
Reducing Overall Consumption		44.35%	.70
spending less money on buying things	.76		
holding onto products longer (e.g., waiting longer to replace TV, phone, or cars)	.65		
buying less quantity of products	.57		

FL = factory loading; AVE = average variance extracted; CR = compose reliability.

variables that imply a more holistic view of deconsumption (e.g., "Excessive buying is against my religious or political beliefs" or "My religious or political beliefs drive my deconsumption behavior"), instead of specific product choices.

Deconsumption Strategies

The deconsumption strategies load on three main factors. We term the first factor green product search, and the deconsumption strategies contained within it are those that most people likely think of when they consider green products. The second factor, temporal product search, comprises of items that measure people's uses of discarded products, sharing or renting items, and making their own products. Such products generally are not owned or purchased by consumers, as is common in the new sharing economy. Finally, some product strategies focus on spending less money, buying less, and holding on to products longer, that is, reducing overall consumption.

Table 5
Means, Standard Deviations, and Correlations

Variable	M	SD	1	2	3	4	5	6	7	8	9	10	11	12	13	14	15
1. Gender	n.a.	n.a.	---														
2. Age	3.15	1.43	.17**	---													
3. Income	3.27	1.55	.03	.02	---												
4. Ethnicity	n.a.	n.a.	-.06	-.03	.11	---											
5. Education	n.a.	n.a.	.05	.24**	.36**	.01	---										
6. Frugality	3.89	1.02	-.08	-.19**	-.33**	-.01	-.16*	---									
7. Monetary concerns	4.31	.62	-.01	-.18**	-.01	.02	.03	.28**	---								
8. Corporate objections	3.74	.81	.01	-.04	.03	-.10	-.06	.04	-.03	---							
9. Resource concerns	4.08	.62	.15*	-.02	-.06	-.10	-.04	.06	.13*	.54**	---						
10. Simplifying	4.02	.72	.11	.20**	-.05	-.10	-.01	-.02	.27**	.13*	.32**	---					
11. Healthy concerns	3.75	.79	.24**	.10	.05	-.05	-.05	-.12*	.05	.52**	.50**	.30**	---				
12. Community bonding	2.38	.96	-.10	.04	-.02	.03	-.07	.05	-.04	.20**	.12*	.04	.12*	---			
13. Sharing	3.49	.97	-.14*	-.22**	-.15*	-.06	-.12	.07	.09	.26**	.41**	.20**	.23**	.24**	---		
14. Green product	3.69	.65	.02	-.04	-.02	-.07	-.08	-.03	.04	.49**	.49**	.22**	.49**	.23**	.36**	---	
15. Temporal product	3.03	.90	-.01	-.09	-.18**	.01	-.12	.10	.07	.19**	.22**	.22**	.24**	.23**	.53**	.46**	---
16. Reduce overall consumption	4.19	.54	.01	.00	-.12	.05	-.01	.18**	.35**	-.01	.23**	.33**	.16**	-.04	.14*	.29**	.22**

** $p < 0.01$, *$p < 0.05$ (two-tailed); n.a. = not applicable.

Table 5 lists the means, standward deviations, and correlations among all identified latent constructs for deconsumption motivations and strategies.

Links between Deconsumption Motives and Strategies

In this section, we seek to connect the reasons people deconsume with the different product/behavioral strategies they use. Factor items in tables 3 and 4, as well as literature will be used to support the hypothesis. Notably, though we identify religious and political concerns as motives to deconsume, a longitudinal study analyzing nationally representative polls from 1990, 1991, 1999, and 2005 – 2015 suggests that environmental concerns among U.S. Christians have been static or declining, irrespective of their level of religiosity and even after controlling for political orientation (Konisky, 2018). Thus product/behavioral choices based on religion or politics might not be identified by consumers as ways to fulfill their deconsumption motivation. Therefore, we do not believe that individuals driven by these motivations will significantly predict the strategies identified in our study and they will not be part of the remaining discussion. We now propose hypotheses for testing which deconsumption motivations best predict the types of strategies that consumers use.

Consumers labeled green (Jansson et al., 2010) and those interested in healthy items (Prasad et al., 2008) tend to seek products as suggested by the green product search strategy factor (Borin et al., 2013). Their motivations reflect resource concerns and health concerns, as defined previously. In addition, consumers may avoid harmful or non-sustainable products, whether due to their ideological ire against a company or their shared views (Iyer & Muncy, 2009). Some of the green product strategies, such as creating less waste and producing smaller carbon footprints, might be achieved by sharing products with others or not purchasing products from companies not known for responsible activities. We anticipate that the corporate objections, sharing, and community bonding motives reflect this drive. Deconsumption motivations associated with simplifying, frugality, or monetary constraints are unlikely to match a green product strategy though, because greener products are not always cheaper, nor are they generally positioned as a means to declutter or simplify people's lives (Iwata, 1997).

Table 6
Regression Analysis of Deconsumption Strategies

	Green Product Search Model 1	Green Product Search Model 2	Temporal Product Search Model 3	Temporal Product Search Model 4	Reduce Overall Consumption Model 5	Reduce Overall Consumption Model 6
Gender	.03	-.06	.03	.15	.02	-.03
Age	-.02	-.02	-.05	.01	-.00	.00
Income	.01	-.00	-.09*	-.06	-.05	-.04
Ethnicity	0.04	-.01	.01	.03	.03	.04
Education	-.03	-.00	-.02	-.01	.01	.01
Hypothesis 1						
Corporate Objection		.16*				
Resource Concerns		.24**				
Health Concerns		.21**				
Community Bonding		.08*				
Sharing		.08*				
Hypothesis 2						
Simplifying				NS		
Resource Concerns				NS		
Frugality				NS		
Community Bonding				.11*		
Sharing				.48*		
Hypothesis 3						
Monetary Concerns						.23**
Sharing						NS
Resource Concerns						.18**
Simplifying						.16**
Frugality						NS
R^2	.01	.41	.04	.32	.02	.22
Adjusted R^2	-.01	.38	.02	.30	-.00	.20
Change in R^2	.01	.39	.04	.28	.02	.21
F (change)	.73	32.47**	2.15	25.69**	.97	16.41**

** $p < 0.01$, *$p < 0.05$ (two-tailed); NS = non-significant.

H1: The deconsumption motives - resource concerns, health concerns, corporate objections, community bonding, and sharing positively influence a green product search strategy.

In addition, temporal product search strategies should satisfy community bonding and sharing motivations well. People interested in reducing their resource utilization (resource concerns), saving money (frugality), or owning fewer things (simplifying) might use this strategy to deconsume, too (Etzioni, 1998). However, because corporate objections and health concerns reflect motives to deconsume by not buying products made by companies that harm the environment or health, these factors likely do not encourage temporal product search. Furthermore, consumers with monetary constraints might not be able to afford to make their own products or rent other products. Therefore, H2: The deconsumption motives - community bonding, sharing, simplifying, frugality, and resource concerns positively influence the temporal product search strategy.

Reducing overall consumption aligns well with motivations to deconsume due to resource concerns, frugality, monetary concerns, and simplifying (Craig-Lees & Hill, 2002; Etzioni, 1998; Iwata, 1997, 1999). Sharing is another way to reduce consumption by spending less money or buying smaller quantities. Health concerns, corporate objections, and community bonding often lead to purchases to achieve specific deconsumptive outcomes though, so they are unlikely to reduce consumption. We propose:

H3: The deconsumption motivations of sharing, simplifying, frugality, monetary concerns, and resource concerns positively influence the reducing overall consumption strategy.

As mentioned, we used sequential regressions to test these proposed hypotheses.

Regression and Results

With sequential regressions, we then test the relationship among deconsumption motivations and strategies. By calculating the mean of all items comprising each of these factors, we created motivations and strategies variables. Using the method suggested by Cohen et al. (2003), we entered the demographic variables first as control variables (Models 1, 3, and 5), followed by the deconsumption motivations for each strategy (Models 2, 4, and 6). The results are in Table 5. The R-square statistics suggest that the models with deconsumption motivations are significantly better than the demographic models. For each regression equation, the tolerance is greater than .1 suggesting multicollinearity is not a major concern (Bowerman & O'Connell, 1990). See table 6 for details.

Green product strategies.

The overall model for green product search is statistically significant ($F(14,494) = 32.47$, $p < .001$) and offers good predictive power ($R2 = .41$), in general support of H1. All five hypothesized motivations - resource concerns ($\beta = .24$, $p < .001$), health concerns ($\beta = .21$, $p < .001$), corporate objections ($\beta = .16$, $p < .05$), sharing ($\beta = .08$, $p < .05$) and community bonding ($\beta = .08$, $p <.05$) significantly predict green product search. The resource concerns motivation implies that consumers use green products and exhibit their concern for future generations by preserving scarce resources and consuming less. Therefore, it makes sense that this motivation predicts strategies that involve switching to products that use cleaner resources and reduce overall consumption. Health concerns also are significantly predictive of buying green products, which promise less toxic ingredients. People motivated by corporate objections purchase green products, reflecting their desire to support companies that align with their deconsumption values. The significant results for corporate objections also align with prior research that suggests social motives are significant indicators of green product consumption, such as when consumers engage in conspicuous consumption to establish their social identity (Griskevicius et al., 2010). Similar reasoning can explain the significant influence of community bonding for predicting green product strategies.

Temporal product search.

The overall model to test H2 is significant ($F(14,494) = 25.69$, $p < .001$), with good predictive power ($R^2 = .32$). Both sharing ($\beta = .48$, $p < .05$) and community bonding ($\beta = .11$, $p < .05$) motivations positively predict temporal product search. Surprisingly however, resource concerns, simplifying, and frugality do not appear to influence consumers' choice of temporal product searches.

The finding that sharing and community bonding motivations predict the use of temporal product search strategies is notable for practice. Between 2010 and 2013, startups focusing on connecting consumers or businesses to products and services that would otherwise go unused increased by 4% (Needleman & Loten, 2014). Consumers guided by these motivations are not deterred by the additional time required to rent or make their own products. This result also resonates with social identity theory, in that social identification leads people to engage in activities that are congruent and supportive of the institution that it embodies (Ashforth & Mael, 1989). These consumers like recycled and discarded products that can be reused; they are interested in engaging with others in less traditional ways (Griskevicius et al., 2010).

Neither frugality nor resource concerns is significant though, contrary to our expectation. Perhaps renting or making one's own (temporal) products require more costly resources (time, money) in the long term, reducing their attraction for consumers motivated by saving money, as well as for those who seek "safer" versions of a product. The simplifying motivation also does not significantly influence temporal product searches. It appears that the simplifying motivation is truly a motivation to lead a simple life, which people achieve by spending and buying less. The motivation to live simply does not significantly predict temporal search strategies such as renting versus buying, making products, or choosing greener versions of products, which could be perceived as additional complications rather than means of simplifying.

Reducing overall consumption.

We find some support for H3 ($F(14,494) = 16.41$, $p < .001$, $R2 = .22$). As hypothesized, this overall consumption reduction strategy is significantly driven by monetary concerns ($\beta = .23$, $p < .001$), simplifying ($\beta = .16$, $p < .001$), and resource concerns ($\beta = .18$, $p < .001$). However, sharing and frugality do not significantly predict a strategy of reducing overall consumption.

People who are motivated to deconsume due to monetary concerns want to spend less money and buy fewer products. Resource concerns also center on caring for future generations by preserving scarce resources and consuming less. Therefore, it makes sense that these motives predict strategies that reduce overall consumption. Not buying items, decluttering one's life, and avoiding materialism stem from simplifying, so this factor also is significant. However, sharing often requires some initial purchase, which may explain why this motivation was not significant. Interestingly, frugality did not significantly predict overall reduction in consumption, a finding that we return to in future research.

Discussion

In a review of two decades (2000 – 2020) of research conducted on sustainable consumption, Quoquab and Mohammad (2020) identify several gaps in the literature pertaining to the lack of definition, dimensions, measures, and practice and policy implications related to the concept. Our study addresses several of these knowledge gaps. For example, the study derives a theoretically grounded and methodologically sound definition of deconsumption, "Consuming products or services uses energy, food, or materials; a deconsumer is a person who has modified his or her consumption to use less energy, food, or materials over time." Further, the study results establish two important findings for the growing field of deconsumption research. First, motivation factors indicate a vast variety of reasons people deconsume, and various stakeholders such as public policy makers, nonprofit organizations, and marketing departments can use these motivations to target their messages effectively. Second, deconsumptive motivations create new opportunities for companies to sell their products, by leveraging their ability to meet some deconsumptive motivational need. Extant research has established that individual consumption significantly impacts overall environmental issues. We believe that the empirical findings of our research give us grounds to propose strategic implications to several stakeholders to reduce their adverse effects on the environment. Although exploratory in nature, we discuss some of these possibilities for stakeholders.

Public policy makers should address the widely acknowledged gap between consumers' positive attitudes toward sustainability and their actual behavior. Forty percent of consumers say they are willing to buy "green" products; only 4% actually do (United Nations Environment Programme, 2005). This gap frustrates producers of sustainable products that rely on the positive predictions and then confront low actual demand (Prothero et al., 2011). Understanding the segments and different motives of various deconsumers can help guide policy that facilitates the success of producers of sustainable products, as well as encourage increased uses of sustainable products. Macro-institutional approaches (public policy, education, and government) to deconsumption should appear in both policy and research. Relying solely on consumers to deconsume by their own choice is not a reasonable expectation (Prothero et al., 2011).

Notably, "Sustainable Consumption and Production Plans" is one of the United Nation's 17 Sustainable Development Goals (https://sustainabledevelopment.un.org/?menu=1300). The overall goal is to focus businesses and consumers on the impact their consumption makes on planetary resources. Governments and policy makers can examine the motivational reasons to consume that we identify herein, then target both products and messages to the public in ways that are likely to lower their consumption. Regulatory decisions related to the new sharing economy could help encourage consumers motivated to deconsume. Publications such as the green guide (goodguide.com) might highlight companies whose products can help consumers deconsume. Groups such as the Sierra Club or other environmental organizations can use their collective efficacy to address deconsumers motivated by community bonding interests.

Many of the products that informed our three deconsumptive product options historically have focused on green features (reducing waste, recycling, using more environmentally safe ingredients) or saving money (holding on to products longer). The results of this study reveal that many companies could benefit by adding messages that appeal to deconsumptive attitudes. As proposed by Ewing, Allen, and Ewing (2012), marketers should create visual and verbal cues to enhance the congruence between consumers' expectations and product features. They suggest cues that emphasize specific product attributes consistent with a consumer's perception of "green" can yield more favorable attitudes toward that product. According to our findings, businesses and nonprofits also could use deconsumptive cues to communicate. For example, a smaller bed may save money; due to its smaller footprint, it also supports smaller houses, which reduce energy demands. Home appliance manufacturers could appeal to a sharing motivation by communicating that the shared aspect of group at-home cooking decreases food consumption overall.

Noting the concept of "green marketing myopia" (Ottman et al., 2006), product managers should investigate product positioning that highlights not just environmental but also other, non-green benefits, such as community bonding, sharing, or corporate alignment. For example, marketers of green and recycled products could use slogans like, "Recycled products strengthen my sense of community," to appeal to the deconsumption motive of community bonding. Heirloom products such Le Creuset pots or Vespas (handed down from one generation to another) could appeal to frugality (spending less money over time for durable products), community bonding, or health (e.g., making one's own food) motives, so an appeal to deconsume could focus on the longevity and beauty of the products.

Limitations and Further Research

We restricted the sample to a U.S. population. Additional research might investigate cultural differences in deconsumption motivations and identify any variance in these motives, as well as in the product choices and strategies that users adopt to enable themselves to deconsume. Replication of the study in different contexts can also refine the measurement items used for this study.

To move beyond the goals of the current project, further research might conduct segmentation analyses according to deconsumption motives, to determine whether the motivations are mutually exclusive and allow businesses to target specific customer groups. It also would be interesting to investigate what role collective efficacy plays among consumers motivated by sharing or bonding and whether their perceptions of collective efficacy drive specific product choices or strategies (Jugert et al., 2016). We also find it interesting that people motivated to consume for religious or political reasons do not think any of the product strategy options would help them deconsume. Perhaps, similar to community bonding, they are more external motivations. Continued work might explicitly investigate the differences between extrinsic and intrinsic motivations. Finally, the non-significant impact of frugality on overall consumption reduction is also warranted. Understanding the difference between frugality and monetary constraint can help bring in the personality aspect to this line of research.

Conclusion

The lifecycle analysis of a product evaluates the environmental impact from materials extraction, manufacturing, packaging, distribution, consumption, and disposal. These activities require resources (e.g., trees for wood) and emit carbon into the atmosphere, which has been linked to climate change. The recent report from the ICCP demonstrates that greenhouse gas emissions continue to rise and that current plans to limit temperatures to 1.5 Celsius above pre-industrial levels may not be enough.

If there is good news, it seems that multiple stakeholders are beginning to understand the seriousness of the issue. Governments are enacting legislature that increases renewable energy and holding companies responsible for gas emissions. Companies are reducing packaging while introducing less environmentally destructive products. A recent study (MasterCard Newsroom, 2021) demonstrates that "58% of adults are more mindful of their impact on the environment, and 85% said they're willing to take personal action to combat environmental and sustainability challenges in 2021", while over half said they believe they must personally reduce their carbon footprint.

As lifecycle analysis shows, consumer product consumption can dramatically impact the energy and other resources used and outputs emitted. The consumer study above also shows that consumers realize this, and many are altering their behaviors. Many studies have examined consumers who are looking into reducing their consumption. Our study demonstrated that there is a multitude of motivations for consumer deconsumption and encouragingly that consumers believe there are many products that can help them achieve this goal. These product strategies will differentially appeal to consumers based on their deconsumption motivations.

Consumers will always buy products, but if companies continue to manufacture or sell less environmentally harmful products (e.g., green products, shared products, reused products), it would be advantageous for both them and the planet to sell their products vis-à-vis harmful products. The deconsumption motivations uncovered in this project can be a new way to do this.

References

Anderson, J. C., & Gerbing, D. W. (1988). Structural equation modeling in practice: A review and recommended two-step approach. *Psychological bulletin*, 103(3), 411-423.

Antil, J. H. (1984). Socially responsible consumers: Profile and implications for public policy. *Journal of macromarketing*, 4(2), 18-39.

Ashforth, B. E., & Mael, F. (1989). Social identity theory and the organization. *Academy of management review*, 14(1), 20-39.

Black, I. R., & Cherrier, H. (2010). Anti-consumption as part of living a sustainable lifestyle: Daily practices, contextual motivations and subjective values. *Journal of Consumer Behaviour*, 9(6), 437-453.

Borin, N., Lindsey-Mullikin, J., & Krishnan, R. (2013). An analysis of consumer reactions to green strategies. *Journal of Product & Brand Management*, 22(2), 118-128.

Bowerman, B. L., & O'connell, R. T. (1990). *Linear statistical models: An applied approach*. 2nd ed. Belmont, CA: Duxbury.

Cohen, J., Cohen, P., West, S. G., & Aiken, L. S. (2003). *Applied multiple correlation/regression analysis for the social sciences*. New York: Psychology Press.

Costello, A. B., & Osborne, J. (2005). Best practices in exploratory factor analysis: Four recommendations for getting the most from your analysis. *Practical assessment, research, and evaluation*, 10(1), 7.

Cowles, D., & Crosby, L. A. (1986). *Measure validation in consumer research: A confirmatory factor analysis of the voluntary simplicity lifestyle scale*. ACR North American Advances.

Craig-Lees, M., & Hill, C. (2002). Understanding voluntary simplifiers. *Psychology & Marketing*, 19(2), 187-210.

Dupré, S. (2005). *Talk the walk: Advancing sustainable lifestyles through marketing and communications*. UNEP/Earthprint.

Durif, F., Boivin, C., Rajaobelina, L., & François-Lecompte, A. (2011). Socially responsible consumers: profile and implications for marketing strategy. *International Review of Business Research Papers*, 7(6), 215-224.

Etzioni, A. (1999). Voluntary simplicity: Characterization, select psychological implications, and societal consequences. *In Essays in Socio-Economics* (pp. 1-26). Springer, Berlin, Heidelberg.

Ewing, D. R., Allen, C. T., & Ewing, R. L. (2012). Authenticity as meaning validation: An empirical investigation of iconic and indexical cues in a context of "green" products. *Journal of consumer behaviour*, 11(5), 381-390.

Fabrigar, L. R., Wegener, D. T., MacCallum, R. C., & Strahan, E. J. (1999). Evaluating the use of exploratory factor analysis in psychological research. *Psychological methods*, 4(3), 272-299.

Fava, J. L., & Velicer, W. F. (1992). The effects of overextraction on factor and component analysis. *Multivariate behavioral research*, 27(3), 387-415.

Fielding, K. S., & Hornsey, M. J. (2016). A social identity analysis of climate change and environmental attitudes and behaviors: Insights and opportunities. *Frontiers in psychology*, 7, 121.

Flatters, P., & Willmott, M. (2009). Understanding the post-recession consumer. *Harvard Business Review*, 87(7/8), 106-112.

Fornell, C., & Larcker, D. F. (1981). Evaluating structural equation models with unobservable variables and measurement error. *Journal of marketing research*, 18(1), 39-50.

Grinstein, A., & Nisan, U. (2009). Demarketing, minorities, and national attachment. *Journal of Marketing*, 73(2), 105-122.

Griskevicius, V., Tybur, J. M., & Van den Bergh, B. (2010). Going green to be seen: status, reputation, and conspicuous conservation. *Journal of personality and social psychology*, 98(3), 392-404.

Hartman, G. (2007). *Hartman report on sustainability: understanding the consumer perspective*. Hartman Group Bellevue, WA.

Haws, K. L., Winterich, K. P., & Naylor, R. W. (2012). *Seeing the world through GREEN-tinted glasses: how green consumers use motivated reasoning to prefer environmentally friendly products*. Unpublished manuscript, Texas A&M University, College Station, Texas, United States.

Iwata, O. (1997). Attitudinal and behavioral correlates of voluntary simplicity lifestyles. *Social Behavior and Personality: an international journal,* 25(3), 233-240.

Iwata, O. (1999). Perceptual and behavioral correlates of voluntary simplicity lifestyles. *Social Behavior and Personality: an international journal,* 27(4), 379-386.

Iyer, R., & Muncy, J. A. (2009). Purpose and object of anti-consumption. *Journal of business research,* 62(2), 160-168.

Jansson, J., Marell, A., & Nordlund, A. (2010). Green consumer behavior: determinants of curtailment and eco-innovation adoption. *Journal of consumer marketing,* 27(4), 358-370.

Johnston, T. C., & Burton, J. B. (2002). Voluntary simplicity: popular definitions and major themes.

In Allied Academies International Conference. Academy of Marketing Studies. Proceedings (Vol. 7, No. 1, p. 27). Jordan Whitney Enterprises, Inc.

Jugert, P., Greenaway, K. H., Barth, M., Büchner, R., Eisentraut, S., & Fritsche, I. (2016). Collective efficacy increases pro-environmental intentions through increasing self-efficacy. *Journal of Environmental Psychology,* 48, 12-23.

Konisky, D. M. (2018). The greening of Christianity? A study of environmental attitudes over time. *Environmental Politics,* 27(2), 267-291.

Lee, M. S., Fernandez, K. V., & Hyman, M. R. (2009). Anti-consumption: An overview and research agenda. *Journal of Business Research,* 62(2), 145-147.

Long, H., Shi, S., Tang, Z., & Zhang, S. (2022). Does living alone increase the consumption of social resources? *Environmental Science and Pollution Research International,* 29(47), 71911-71922.

Marini, M.A., Tarola, O. & Thisse, J.F. (2022). When is Environmentalism Good for the Environment? *Environmental and Resource Economics, 82* 821-828.

Master Card Newsroom. (2021, April 12). *Consumer passion for the environment grows as a result of the pandemic.* https://www.mastercard.com/news/insights/2021/consumer-attitudes- environment/.

Mohr, L. A., Webb, D. J., & Harris, K. E. (2001). Do consumers expect companies to be socially responsible? The impact of corporate social responsibility on buying behavior. *Journal of Consumer affairs, 35*(1), 45-72.

Needleman, S. E., & Loten, A. (2014). Startups want to be the next Airbnb, Uber. *The Wall Street Journal, 7.*

ORCI [Ontario Responsible Consumption Index] (2012) https://commerce.eduzone.ca/en/shop/products-by-subject/case- studies/organizationaldevelopment/ontario-responsible-consumption-index-2012-2/

Ottman, J. A. (2017). The new rules of green marketing: Strategies, tools, and inspiration for sustainable branding. Routledge.

Ottman, J. A., Stafford, E. R., & Hartman, C. L. (2006). Avoiding green marketing myopia: Ways to improve consumer appeal for environmentally preferable products. *Environment: science and policy for sustainable development, 48*(5), 22-36.

Piligrimienė, Ž., Žukauskaitė, A., Korzilius, H., Banytė, J., & Dovalienė, A. (2020). Internal and external determinants of consumer engagement in sustainable consumption. *Sustainability, 12*(4), 1349.

Prasad, A., Strijnev, A., & Zhang, Q. (2008). What can grocery basket data tell us about health consciousness? *International Journal of Research in Marketing, 25*(4), 301-309.

Prothero, A., Dobscha, S., Freund, J., Kilbourne, W. E., Luchs, M. G., Ozanne, L. K., & Thøgersen, J. (2011).

Quoquab, F., & Mohammad, J. (2020). A review of sustainable consumption (2000 to 2020): What we know and what we need to know. *Journal of Global Marketing, 33*(5), 305-334.

Roper, A. (2002). Green Gauge Report. https://www.gfk.com/gfk-green-gauge

Sustainable consumption: Opportunities for consumer research and public policy. *Journal of Public Policy & Marketing, 30*(1), 31-38.

Sheth, J. N., Sethia, N. K., & Srinivas, S. (2011). Mindful consumption: a customer-centric approach to sustainability. *Journal of the academy of marketing science, 39*(1), 21-39.

Sun, Y., Tieshan, L., & Wang, S. (2022). "I buy green products for my benefits or yours": understanding consumers' intention to purchase green products. *Asian Pacific Journal of Marketing and Logistics, 34(8),* 1721-1739.

United Nations. (2022). *The 17 Goals.* https://sdgs.un.org/goals#goals

Webb, D. J., Mohr, L. A., & Harris, K. E. (2008). A re-examination of socially responsible consumption and its measurement. *Journal of business research, 61*(2), 91-98.

Zabkar, V., & Hosta, M. (2013). Willingness to act and environmentally conscious consumer behaviour: can prosocial status perceptions help overcome the gap? *International Journal of Consumer Studies, 37*(3), 257-264.

The California State University Journal of Sustainability and Climate Change
2022, Volume 2

Energy and Climate Change Issues Around CSUDH

Alex (Adriana) E. Perez (California State University, Dominguez Hills), Tara B. Jones (California State University, Dominguez Hills), Raju Bista (California State University, Dominguez Hills), Parveen K. Chhetri (California State University, Dominguez Hills)

Abstract

Climate change is posing significant challenges to California's energy sector. Extreme weather events (heat and cold) may pressure existing infrastructure. Many studies have indicated that extreme climate events would impact the energy system by affecting peak electricity demand. However, very few studies have been conducted to understand how disadvantaged communities (DACs) will be impacted. Because of unequal access to energy infrastructure (electricity generation and battery storage), DACs are more vulnerable to power outages due to the rising number of significant weather events caused by climate change. To address the issue of how DACs are disproportionately affected by climate change-related energy issues, we used DACs, infrastructure, and climate data. We identified the most vulnerable communities associated with climate change-related energy issues in areas around California State University Dominguez Hills (CSUDH). This study's findings will support building the resilience of energy infrastructure to climate change and minimize the energy burden on DACs.

Introduction

Climate is changing worldwide, and extreme weather events are increasing. The recently released Intergovernmental Panel on Climate Change (IPCC) report concludes that the average global temperature is likely to rise 1.5 °C, or 2.7 °F, above preindustrial levels by 2040 (IPCC, 2021). The report also indicated that at that threshold, nearly 1 billion people could face life-threatening heat waves at least once every five years. Extreme weather events are the leading cause of power outages and a constant hazard to the nation's energy system (Reidmiller et al., 2018). Even though the energy sector is one of the most resilient United States (US) economic sectors, climate change is likely to pose considerable new challenges to this sector (Vine, 2012). Due to climate change, future extreme events that can cause power outages are projected to be more frequent and last longer (USGCRP, 2018). However, existing US energy infrastructures are not designed to operate under extreme weather events, and changes in climate have the potential to cause significant impacts (Zamuda et al., 2013). For example, heat waves increase energy demand from customers and can put pressure on the electric grid and result in grid failure.

It is important to note that climate change does not affect everyone similarly. However, very few studies have been conducted to understand how disadvantaged communities (DACs) will be impacted. These communities experience heightened risk and increased sensitivity to climate change and have less capacity and fewer resources to cope with, adapt to, or recover from climate impacts (CPUC, 2021). These disproportionate effects are caused by physical (built and environmental), social, political, and economic factors exacerbated by climate impacts. DACs are particularly vulnerable to power outages resulting from the rising number of significant weather events under changing climate. Decades-long pervasive socio-economic conditions—perpetuated by systems of inequitable power and resource distribution—are such that they may lack the financial and organizational resources to respond to and recover from climate disasters like drought, flooding, fires, and heatwaves (CNRA, 2017; Roos, 2018).

The increasing reliability of California's electricity sector for performing essential daily activities makes it vulnerable to climate change. Californians are facing three main challenges as a result – an increase in energy demand, the ability of the electricity generation system to adapt, and risk to transmission and distribution networks (Vine, 2008; 2012). Due to climate change, future building space cooling and heating demands could increase significantly. For example, a slight increase in mean temperature can result in a significant rise in Cooling Degree Days (CDD), which could lead to an increase in cooling energy and increase in heating energy (Heating Degree Days -HDD) in case of an extended cold period (Mutschler et al., 2021). Many studies (Zamuda et al., 2013; Wang & Chen, 2014; Burillo et al., 2019; Mutschler et al., 2021) indicated that extreme climate events would impact the energy system by affecting peak electricity demand. In one of the earliest studies on the effects of climate change on the California electricity sector, Baxter and Calandri (1992) indicated that increased demand for cooling may substantially outweigh the heating needs (Vine, 2012). This argument by Baxter and Calandri (1992) makes sense because Franco and Sanstad (2006) found a high correlation between the simple average daily temperature and daily peak electricity demand in California. Peak electricity demand of Los Angeles County residential and commercial sectors was projected to increase from 9.5-12.8 GW in 2010 to 12.3-16.7 GW (~30%) by 2040 and to 13.1-19.2 GW (~45%) by 2060 (Burillo et al., 2019). Extreme weather events will lead to more frequent use and additional installations of cooling and heating units in residential and business buildings. This will significantly increase the peak demands, and the probability of blackouts will increase during the peak demand periods. Therefore, greater climate-resilient energy is needed, and it will require improved technologies, policies, information, andstakeholder engagement (Zamuda et al., 2013).

To further understand how climate change may impact DACs' energy-related issues, we asked two questions:

- Is current infrastructure capable of future energy demand triggered by climate change?
- Are disadvantaged communities more vulnerable to climate change because of poor energy infrastructure?

To address these research questions, we:

- Identified the most vulnerable communities associated with climate change-related energy issues
- Assessed the grid reliability issues (Integration Capacity Analysis) around the CSUDH campus
- Determined critical physical infrastructure assets that are vulnerable or susceptible to failure under different climate change scenarios

Methods

Study Area

We focused our study around California State University Dominguez Hills (CSUDH) campus. CSUDH is a highly diverse, metropolitan university primarily serving the South Bay area of Los Angeles (LA) County (Figure 1). The current student population is around 17,000, and CSUDH is one of the rapidly growing CSU campuses. CSUDH is located in Carson and is surrounded by cities like Compton, Long Beach, Gardena, and Torrance. The majority of communities around CSUDH are classified as DACs.

Identifying Most Vulnerable Communities

Identifying and mapping vulnerable communities associated with climate change-related energy issues is an essential part of the scientific foundation for understanding the state's changing conditions related to climate change (Roos, 2018). There are several indicators developed to identify the spatial pattern of vulnerable communities. We used Environmental Justice Cumulative Impact and CalEnviroScreen Scores to identify DACs and communities most vulnerable to climate change. Environmental Justice Screening Method (EJSM) was developed by USC / Occidental College (Hoffman, 2022). This was initiated by the proposed Green Zones Program at the Department of Regional Planning. California's Environmental Protection Agency (CalEPA) created a screening process known as "CalEnviroScreen 4.0." This tool determines which communities are disadvantaged based on detailed census data and 21 individual indicators (OEHHA, 2021). These 21 indicators are then combined and calculated to create an overall score known as the Cumulative Impact Score (CIScore).

SCE Resources

Southern California Edison's (SCE's) Integration Capacity Analysis (ICA) User Guide (updated 8/30/21) describes the uniform generation-static grid integration capacity, published to its Distribution Resources Plan External Portal (DRPEP), as being the final ICA result based on the most limiting power system criteria at the most limiting hour (SCE, 2021). In the ICA map, red colors represent areas of low integration capacity, and green represents higher integration capacity. SCE produced and made available the 576 hourly ICA values using a "technology-agnostic uniform generation and uniform load" approach, which generates ICA values independent of the type of Distribution Energy Resource (DER) technology. The 576-hourly profile is the monthly minimum and peak load that occurs during each of the 24 hours in a day (Stanfield et al., 2021).

Our approach overlaid the ICA results for uniform generation static grid (Megawatt - MW) with DACs, as defined by CalEnviroScreen, to examine whether the lowest integration capacity result of 100 kW or less was correlated with DACs in our study area. To visualize the ICA results for a wider area than the limitation of a 0.6-mile elevation imposed by SCE in viewing the data, the ICA – Circuit Segments dataset published by the SCE C-GIS Project was opened in ArcGIS Pro 2.8.3 and clipped with the geoprocessing tool to the Los Angeles County boundary. Following the guidelines laid out by Stanfield et al. (2021) regarding the conservative nature of values that account for operational flexibility, we opted to use uniform generation without operational flexibility in our study.

Figure 1

California State University Dominguez Hills (CSUDH). Note: Inset map (upper left) shows 23 CSU campuses, and the inset map (upper right) shows the location of CSUDH in southern California.

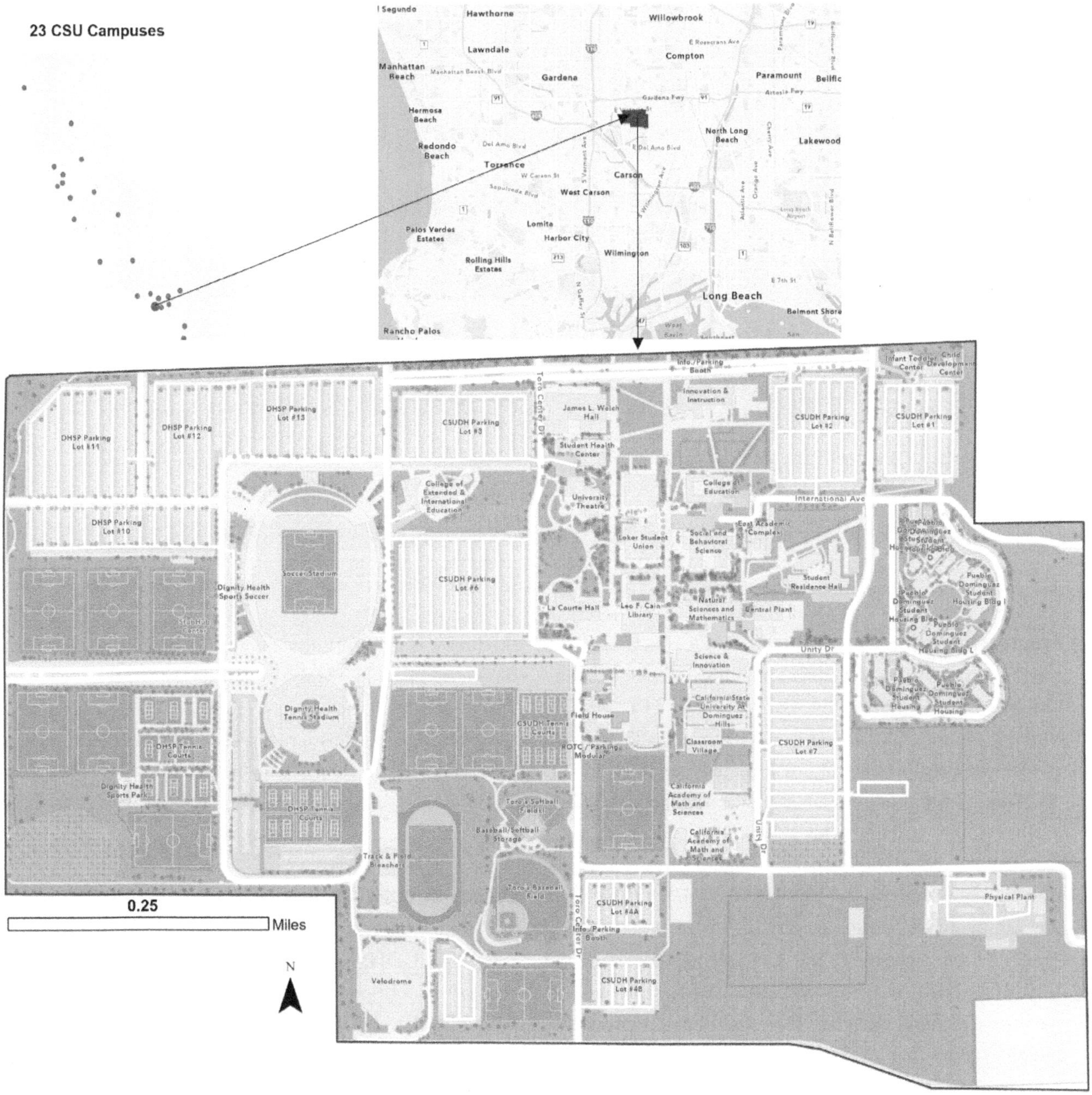

Figure 2
Disadvantaged Communities (DACs) based on Environmental Justice CI Score Surrounding CSUDH Campus

Identify Infrastructure Status in DACs

To assess the infrastructures around CSUDH for climate change-related vulnerability, we obtained Countywide Building Outlines (2014) GIS data layer from the LA County eGIS Homepage (County of Los Angeles Open Data, 2021). The Countywide building outline dataset contains building outlines (over 3 million) for all buildings in Los Angeles County, including building height, building area, and parcel number (also known as building footprints). This data was captured from stereo imagery as part of the LARIAC2 Project (2008 acquisition) and was updated as part of the LARIAC4 (2014) imagery acquisition. Based on this dataset, we separated buildings older than 1970. We assumed that buildings older than 1970 are less likely to have cooling or heating units and lack the infrastructure that could support the future installation of cooling or heating units without a significant upgrade. We also obtained electric vehicle charging station data from the Alternative Fuels Data Center (AFDC) to analyze the spatial pattern distribution and association with DACs (US Department of Energy, 2021).

Results

Disadvantaged Communities around CSUDH

We investigated the spatial distribution pattern of DACs around CSUDH based on the Environmental Justice CI Score. A majority of the communities surrounding CSUDH fall under the high Environmental Justice CI Score (Figure 2). A high CI Score indicated the vulnerability of these communities to climate change.

Integration Capacity Analysis around CSUDH

The following map depicts ICA values updated on October 21, 2021, for the areas surrounding CSUDH (Figure 3). CSUDH's campus falls within the large number of DACs running through central LA County. Areas of low integration capacity of less than 100 kW uniform generation static grid are depicted by the red lines (Figure 3). Many low-capacity areas exist in the non-DAC regions depicted on the map (Figure 3). However, there are also lines of higher integration capacity designated by green within the DAC. Therefore, a correlation between disadvantaged communities and an ICA result of less than or equal to 100 kW for a uniform generation static grid is not visually apparent in the selected study area.

Infrastructure Conditions Around CSUDH

Infrastructures are very important for developing resilience to climate change. People living in substandard infrastructure are more vulnerable to climate change than people living in modern infrastructure. Figure 4 shows the building age (residential and commercial) around CSUDH, and the majority of the buildings were constructed before 1970. In case of extreme weather conditions, people living in these areas will need to take shelter in cooling and warming centers which are sparsely distributed around the Los Angeles area (Figure 5), and there are accessibility issues because of transportation and distance from home.

Inequality in infrastructure-related issues can be seen in the discrepancy in electric vehicle (EV) charging stations around LA County (Figure 6). The discrepancy shown in the map (Figure 6) is the high density of the electric vehicle charging stations in LA County service planning areas 4 (Metro) and 5 (West). The Metro area is home to the Crypto.com Arena and other tourist attractions, such as the Walt Disney Concert Hall and popular museums. The cluster of available charging stations accommodates those visiting the downtown area.

Figure 3

Integration Capacity Analysis (ICA) around CSUDH. Note: Red lines in the map indicate low hosting capacity and pink polygons on the background represent disadvantaged communities (DACs).

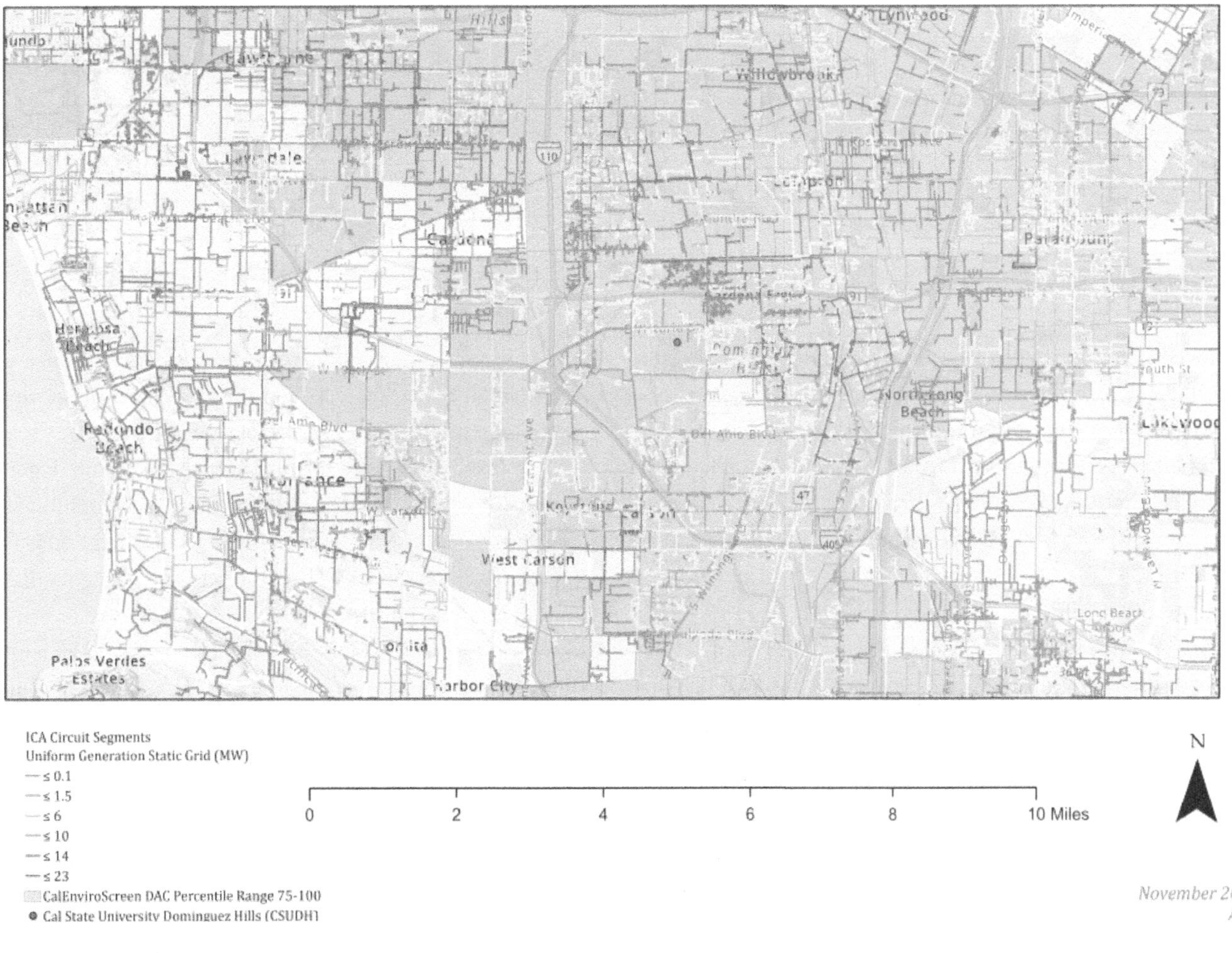

Figure 4

Building Age around CSUDH. Note: Green polygons indicates buildings older than 1970.

However, it is apparent that service planning area 6, the area with the highest population of households paying greater than 50% of their income towards their housing costs, also has the least adequate distribution of EV charging stations (Figure 6).

Discussion

As seen in all the maps presented in this study, there is a clear need for action for the communities around the CSUDH area. Areas that have been historically identified as marginalized are visibly lacking the proper infrastructure that is meant to support the transition to cleaner energy systems. In the map shown in Figure 2, there is a visual correlation between the areas ranking highest with the Environmental Justice CI score, and in Figure 3 where we see communities ranking highest with the CalEnviroScreen CI score. This visual correlation, alongside the data presented in Figure 6, displays the concentrations of communities that we hope policymakers will focus on moving forward to minimize the impact of climate change.

The LA Service Area Planning Map shown in Figure 6 indicated that zone 6 has the highest concentration of households in all of LA County paying greater than 50% of their income towards their housing costs, which identifies them as housing burdened (OEHHA, 2021). That being said, we can see why most households cannot afford climate-resilient energy upgrades. With this data, we can assume that few families can afford to install solar panels and that most of these homes in service planning area 6—older than 1970—are less likely to have cooling or heating units. These disproportionate effects are caused by physical (built and environmental), social, political, and economic factors exacerbated by climate impacts (CPUC, 2021).

Figure 3 shows how strained the current hosting capacity is of the existing static grid, meaning that any upgrade that does take place from homeowners who can afford to do so will be adding to the current demand. This can potentially lead to more frequent blackouts as climate change becomes more unpredictable. Disadvantaged communities (DACs) that cannot adapt to climate change will be the hardest hit by extreme temperatures (Roos, 2018). Several studies (Shonkoff et al., 2012; English et al., 2013) have discussed climate change-related issues in DACs. Shonkoff et al. (2012) have argued that health impacts from climate change will disproportionately affect minority populations and low-income neighborhoods and have made the explicit link between environmental justice and climate health hazards. Former California Governor Jerry Brown signed into law Senate Bill 535 (SB 535), requiring a minimum of 10% of the potential revenue (estimated to be up to $1 billion) generated by the cap-and-trade program to be directed to disadvantaged communities to reduce pollution and develop clean energy to adapt to changing climate (English et al., 2013).

When we think about the impacts that SB 535 will have, we also need to consider the demographics of the communities. English et al. (2013) combined climate change population vulnerability scores with environmental justice scores to create Cumulative Impacts Plus Climate Change Vulnerability Scores. This score also indicated that communities surrounding CSUDH fall under high-risk areas. African Americans (46%) and Latinos (36%) reside in the two highest-risk categories (climate change population vulnerability scores of 4 or 5), while 30% of Whites live in these high-risk census tracts. African Americans were almost four times more likely than Whites to reside in census tracts ranked with the highest vulnerability

Figure 5

Cooling and Warming Centers around Los Angeles County, CA. Note: Red polygons on the background represent disadvantaged communities (DACs).

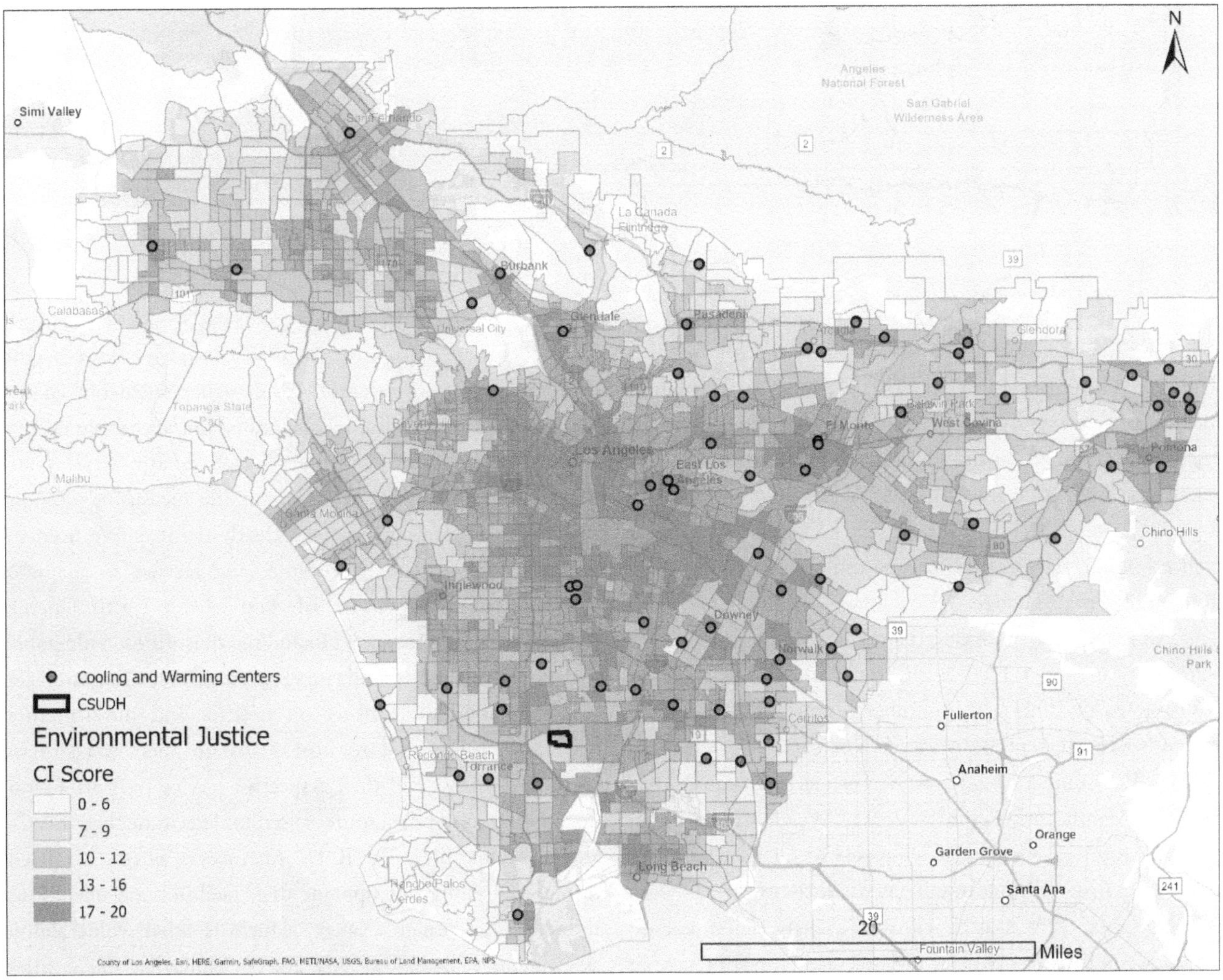

Figure 6

CalEnviroScreen 4.0 Housing Burden Percentage by Census Tract in Los Angeles County Service Planning Areas Relative to Electric Vehicle (EV) Charging Stations

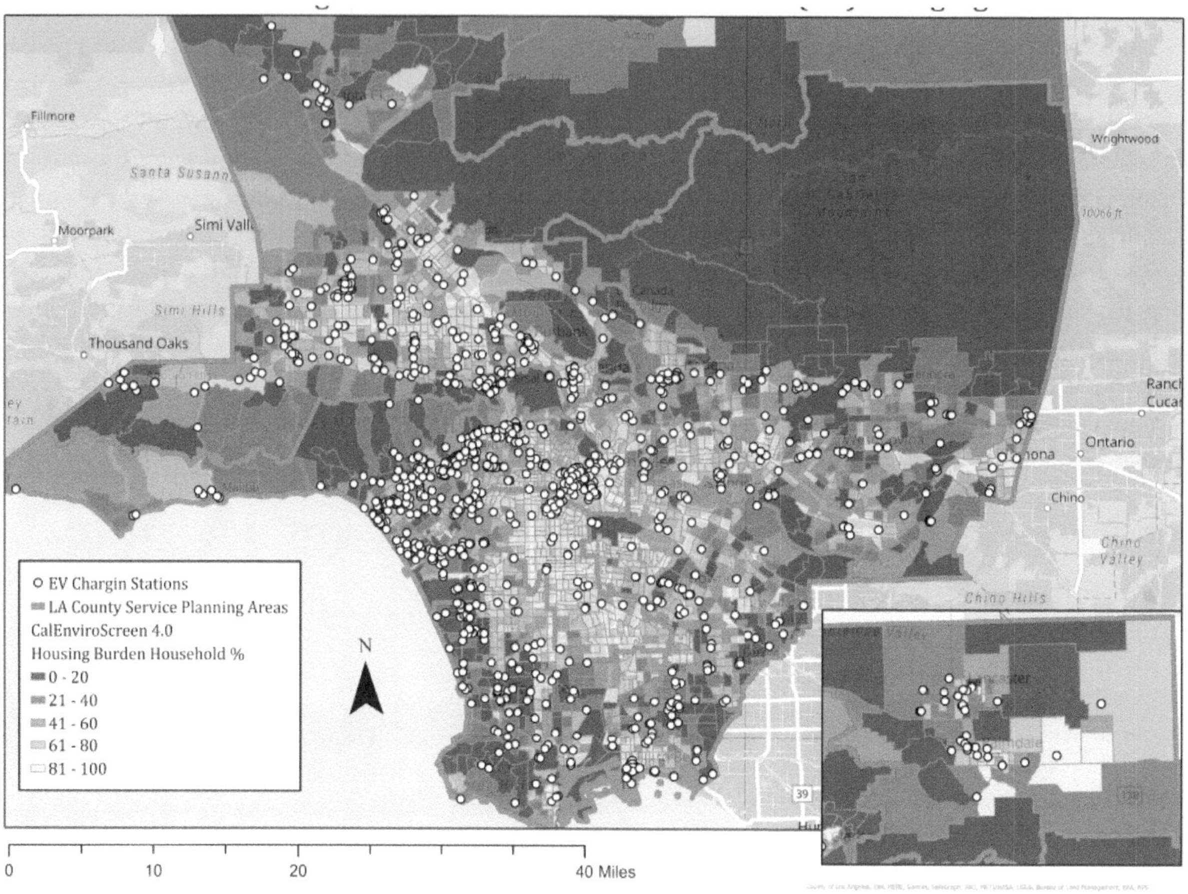

than the lowest vulnerability; Latinos were almost twice as likely as Whites (English et al., 2013). People in the highest risk category were 44% more likely to have emergency room visits for heat illness during a heatwave than in the lowest risk category (English et al., 2013). Most communities around CSUDH are in the high-risk category because of their racial and economic status. These communities will face the extra burden of climate change.

A recent Los Angeles Times study indicated that extreme heat caused 3900 deaths in California over a decade (Wigglesworth, 2021). This number is six times higher than the official count of heat-related fatalities (Wigglesworth, 2021). The number of extreme weather-related fatalities has increased significantly in the last decade. We must identify the spatial pattern of extreme weather-related fatalities and the infrastructure conditions to mitigate these disasters.

To assist DACs with climate change-related energy issues, we must adopt policies and practices protecting people from energy insecurity. Recent policy initiatives have begun to highlight the apparent need to generate energy equity and justice by providing reliable, safe, and affordable energy where the costs and benefits of such energy services are disseminated fairly. However, the ways for measuring progress towards these goals are not yet clearly defined. We need to upgrade infrastructure, operations, and services to adapt to climate change and ensure safe and reliable electric energy service to all Californians, including those most vulnerable and disadvantaged (Disadvantaged Vulnerable Communities - DVCs). Besides focusing on policies and infrastructure upgrades, IOUs need to host maps on their websites to identify the DVCs in their respective service territories with documentation of data sources used to determine these DVCs and all source files. The IOUs must assess where they need to replace, remove, or upgrade their facilities and operations to adapt to climate change, which is what vulnerability assessments will do (CPUC, 2021). Moreover, the utilities need to consider green and sustainable remedies for the vulnerable infrastructure identified in assessing mitigation measures in their vulnerability assessments.

Limitations of this study include Integration Capacity Analysis (ICA) data validation, an issue carried over into the new Order Instituting Rulemaking (OIR). Hosting capacity, by its nature, needs to be continually updated to be useful. Although the IOUs were ordered to update user guides when map functionality changes in a ruling issued 1/27/21 in the R.14-08-013 proceeding, our study identified at least one instance of a user guide not being updated. The public's access to validated energy data will facilitate a smooth transition to a clean energy future. Programs to replace inefficient and natural gas-powered appliances are also necessary to support the transition to clean energy for everyone. Another limitation is that the data does not differentiate between renters, multi-family homes, or homeowners. Future research should attempt to examine how this might affect climate-resilient energy upgrades.

Conclusions

Communities surrounding the CSUDH are economically unprepared for energy-related issues exacerbated by climate change and lack of adequate infrastructure. Because of socioeconomic vulnerability, DACs will have difficulty adapting to climate change. To promote equity, Investor Owned Utilities (IOUs) should conduct extra outreach and education activities in DACs. Future research should examine climate change's impact on energy insecurity.

References

Baxter, L. W., & Calandri, K. (1992) Global warming and electricity demand: A study of California. *Energy Policy*, 20 (3), 233–244. https://doi.org/10.1016/0301-4215(92)90081-C

Burillo, D., Chester, M. V., Pincetl, S., Fournier, E. D., & Reyna, J. (2019). Forecasting peak electricity demand for Los Angeles considering higher air temperatures due to climate change. *Applied Energy*, 236, 1-9. https://doi.org/10.1016/j.apenergy.2018.11.039

California Natural Resources Agency (CNRA, 2017). Safeguarding California: Implementation Action Plans. Sacramento, CA: California Natural Resources Agency (CNRA).

County of Los Angeles Open Data (2021). *County of Los Angeles Open Data* https://data.lacounty.gov/

California Public Utilities Commission (CPUC, 2021). Decision on Energy Utility Climate Change Vulnerability Assessments and Climate Adaptation in Disadvantage Communities. Available online at: https://docs.cpuc.ca.gov/PublishedDocs/Published/G000/M345/K697/345697117.docx

English, P., Richardson, M., Morello-Frosh, R., Pastor, M., Sadd, J., King, G., Jesdale, W., & Jerrett, M. (2013). Racial and income disparities in relation to a proposed climate change vulnerability screening method for California. *The International Journal of Climate Change: Impacts and Responses,* 4(2), 1–18. https://doi.org/10.18848/1835-7156/CGP/v04i02/37156

Franco, G., & Sanstad, A. (2006). *Climate change and electricity demand in California.* CEC-500-2005-2001-SF. California Energy Commission, Sacramento, California.

Hoffman, D. (2022). *EJSM Scores.* Hub.arcgis.com. Retrieved 15 August 2021, from https://hub.arcgis.com/datasets/lacounty::ejsm-scores/about.

IPCC (2021). The Intergovernmental Panel on Climate Change (IPCC), AR6 Climate Change 2021: The Physical Science Basis — IPCC

Mutschler, R., Rüdisüli, M., Heer, P., & Eggimann, S. (2021). Benchmarking cooling and heating energy demands considering climate change, population growth and cooling device uptake. *Applied Energy*, 288, 116636. https://doi.org/10.1016/j.apenergy.2021.116636

Office of Environmental Health Hazard Assessment (OEHHA, 2021). CalEnviroScreen 4.0. OEHHA, California Environmental Protection Agency. Available online at: https://oehha.ca.gov/media/downloads/calenviroscreen/report/calenviroscreen40reportf2021.pdf

Reidmiller, D. R., Avery, C. W., Easterling, D. R., Kunkel, K. E., Lewis, K. L. M., Maycock, T. K., & Stewart, B. C. (2019). Fourth national climate assessment. Volume II: Impacts, Risks, and Adaptation in the United States, Report-in-Brief.Roos, M. (E4 Strategic Solutions) (2018). *Climate Justice Summary Report.* California's Fourth Climate Change Assessment. Publication number: SUM-CCCA4-2018-012.

Roos, M. (2018). Climate Justice Summary Report. California's Fourth Climate Change Assessment. Publication number: SUM-CCC4A-2018-012.

Shonkoff, J. P., Garner, A. S., Siegel, B. S., Dobbins, M. I., Earls, M. F., McGuinn, L., ... & Committee on Early Childhood, Adoption, and Dependent Care. (2012). The lifelong effects of early childhood adversity and toxic stress. *Pediatrics*, 129(1), e232-e246.

Southern California Edison (SCE, 2021). Distribution Resources Plan External Portal. Southern California Edison.

Stanfield, S., Zakai, Y., & McKerley, M. (2021). Key Decisions for Hosting Capacity Analyses (p. 37). IREC. https://irecusa.org/wp-content/uploads/2021/10/IREC-Key-Decisions-for-HCA.pdf

U.S. Department of Energy (2021). Alternative Fuels Data Center. https://afdc.energy.gov/data_download/

U.S. Global Change Research Program (USGCRP, 2018). Impacts, risks, and adaptation in the United States: Fourth National Climate Assessment, Volume II [Reidmiller, D.R., C.W. Avery, D.R. Easterling, K.E. Kunkel, K.L.M. Lewis, T.K. Maycock, and B.C. Stewart (eds.)]. U.S. Global Change Research Program, Washington, DC, USA, 1515 pp. doi: 10.7930/NCA4.2018.

Vine, E. (2008). *Adaptation of California's Electricity Sector to Climate Change*. Public Policy Institute of California (PPIC), San Francisco, CA. http://www.ppic.org/content/pubs/report/R_1108EVR.pdf

Vine, E. (2012). Adaptation of California's electricity sector to climate change. *Climatic Change*, 111(1), 75-99. https://doi.org/10.1007/s10584-011-0242-2

Wang, H., & Chen, Q. (2014). Impact of climate change heating and cooling energy use in buildings in the United States. *Energy and Buildings*, 82, 428-436.

Wigglesworth, A. (2021). California considers ranking heat waves after Times investigation into extreme heart deaths. *Los Angeles Times*. https://www.latimes.com/california/story/2021-11-12/heat-wave-ranking-system-proposed-for-california#:~:text=The%20announcement%20comes%20after%20a%20Times%20investigation%20revealed,a%20lack%20of%20urgency%20in%20confronting%20the%20crisis

Zamuda, C., Mignone, B., Bilello, D., Hallett, K. C., Lee, C., Macknick, J., Newmark, R., & Steinberg, D. (2013). *US energy sector vulnerabilities to climate change and extreme weather*. Department of Energy Washington DC. https://apps.dtic.mil/sti/citations/ADA583709

The California State University Journal of
Sustainability and Climate Change
2022, Volume 2

Post Wildfire Vegetation Response to the Wildland-Urban Interface: A Case Study of the Station Fire

Angelo De Guzman (California State University, Dominguez Hills), Raju Bista (California State University, Dominguez Hills), and Parveen Chhetri (California State University, Dominguez Hills)

Abstract

In the past, wildfires served as a method for mother nature to promote biodiversity and to help maintain a functioning ecosystem. However, climate change alters the fire regime, significantly impacting vegetation recovery. Human disturbances and increased land use and land cover heighten vegetation disruption and abundance after a fire. Wildland-urban interface (WUI) – the region where the vegetation intermingles with the roads, houses, and human-made structures – threatens vegetation and the human population. Overall vegetation recovery after the Station Fire of 2009 spread through the San Gabriel Mountains, Los Angeles County was observed using Digital Elevation Model (DEM), Normalized Difference Vegetation Index (NDVI), and Normalized Difference Burn Ratio (nDBR) spectral indices. In addition, Light Detection and Ranging (LiDAR) images were used to measure aboveground biomass (AGB). The study analyzed vegetation biomass recovery by comparing human disturbances and the level of fire severity within the Station Fire perimeter. Low and moderate fire severity were compared in detail against WUI and non-WUI regions by quantifying the amount of biomass in the specified regions. Linear regression model results showed vegetation recovery rates were slower in WUI regions than in non-WUI regions despite having similar regeneration patterns while AGB rebound was similar across both region categories.

Introduction

Wildfires are important natural disturbances that help maintain the equilibrium of chaparral ecosystems. Fire can rid the natural ground cover of diseases and nourish the soil. Natural fire regimes facilitate the recovery of certain vegetation types. The disruptions in the environment caused by wildfires can provide a quick opportunity for environmental renewal. However, anthropogenic factors such as climate change-induced extreme fire hazard weather patterns, human-related ignition sources, and land abandonment practices have all increased the prevalence and the risk of uncontrolled fires (Martín-Alcón et al., 2015). Wildfire amplification results in vegetation loss, soil disturbance, and a reduction in biological activity (Efthimiou et al., 2020).

In the chaparral ecosystem, biomass can be rapidly replenished by long-interval fires. A typical chaparral landscape is shrub-dominated and is usually found at medium elevations. After a wildfire, the chaparral shrubland will promote seed production to stimulate rapid regeneration. The resilience of this vegetation type enables it to flourish after a wildfire. There is only one caveat: if there are repeated or multiple fires over short periods at the same site, the chaparral will not be able to adapt and regenerate (Storey et al., 2021). A change in fire intensity or frequency can thus profoundly impact the chaparral ecosystem's structure and composition over time (Barro & Conrad, 1991). For example, increased wildfire frequency or variation, can reduce seed dormancy periods and promote seed mortality.

Aside from wildfires, human disturbances also play a vital role in reducing vegetation biomass. The expansion of urban sprawl over recent decades has led to the intensification of the wildland-urban interface (WUI), which has severely impacted chaparral ecosystems. The WUI region refers to areas in which human-made structures intermingle with wild vegetation and increased human activity in WUI regions can greatly exacerbate the fire risk (Syphard et al., 2021). Human disturbances can be classified in various ways, including recreational activities, consumptive and non-consumptive activities like camping, off-roading, gardening, cooking, casual outdoor activities, picnicking, and hiking. Human disturbances also refer to aspects of the built environment, such as buildings and roads. The expansion of the WUI puts not only the chaparral ecosystems at risk, but the human-made infrastructure at risk as well.

With an ever-growing population, buildings and infrastructure construction programs will continue to expand. As a result, there is an amplified interaction between humans and the environment in both WUI and non-WUI areas. It is no coincidence that wildfires in Southern California are estimated to be rising exponentially in frequency and intensity. It is also noteworthy that human activities are major factors in the decline of vegetation rebound capacities, especially in high-mountainous locations (Rull et al., 2011). Human-induced climate change, and human impacts on land cover and land use, all negatively impact the chaparral and similar vegetation types in California. As a result of climate change, droughts are increasing, slowing down vegetation growth.

Surface topography is another significant factor affecting vegetation recovery. There is a particular pattern of burning that intensifies and lengthens wildfires due to elevation, aspect, and soil type (Keeley & Safford, 2016). Wildfires in elevated zones and sloped areas change soil properties, including permeability (Efthimiou et al., 2020). As a result, the soil becomes hydrophobic, reducing soil absorption and creating an inhabitable layer for vegetation. Thus, repeated burnings and human disturbances in higher elevation and steeper slope areas particularly impede natural vegetative recovery.

Satellite-based monitoring technologies can now be used to accurately quantify vegetation biomass. The Normalized Difference Vegetation Index (NDVI) and Normalized Difference Burn Ratio (nDBR), both generated from remote sensing imagery, can delineate fire patterns, characterize burn severity, and identify changes in vegetation structure (McCarley et al., 2017; Wulder et al., 2009). Satellite-based monitoring quantifies NDVI to approximate vegetation biomass, crown and canopy cover, vegetation density, and vegetation health. Satellite-based imagery from Landsat and Sentinel-2 are popularly used to capture high-resolution images to calculate NDVI. Basically, NDVI estimates vegetation by measuring the difference between the near-infrared and red wavelengths of the electromagnetic spectrum. Vegetation strongly reflects near-infrared while absorbing red wavelength in contrast. A potential indicator of burn severity could be used to manage the functionality of a given burned area as a measure of ecosystem rehabilitation. Similar to NDVI, the nDBR spectral index identifies the burned areas of the vegetation. The ratio between the near-infrared and short-wave infrared wavelengths are used to calculate the nDBR of a given area. The nDBR estimates the burn severity of a targeted area to specify severe vegetation damage or regrowth post-wildfire.

Three-dimensional remote sensing models such as Light Detection and Ranging (LiDAR) are other beneficial tools for digitally modelling vegetation biomass. LiDAR is an active remote sensing tool that utilizes sensors and returned pulses to categorize various classified points such as ground level, low, medium, or high vegetation. It involves mounting a laser instrument on a low-flying aircraft that emits pulses of light that reflect off of objects and then back to the sensor. LiDAR is an accurate tool for reconstructing, recreating, and measuring three-dimensional vegetation structures (Guo et al., 2017). LiDAR has thus emerged as a powerful tool for characterizing post-disturbance regeneration assessments to accurately characterize vegetation attributes (Martín-Alcón et al., 2015).

Two major concerns that pose a threat to vegetation biomass post-wildfire are short-interval fires and human disturbances. Daily recreational activities and fuel sources provided by humans can increase fire risk and fire prevalence. Short-interval fires and repeated burning at the same locations can severely impact the topography of the land, fire regime, and soil regime. However, it remains unknown the degree to which short-interval fires destroy or inhibit overall vegetation biomass regrowth. This study aimed to explore the recovery of vegetation biomass post-wildfire in the presence of human activity in a WUI area outside Los Angeles, California. It focused on addressing the question of whether wildfires and the WUI play a ubiquitous role in limiting post-wildfire vegetation recovery. Taking a broader, landscape-wide approach, the objectives of the study were to (1) explore visual accounts of the land cover and land use in the WUI and non-WUI regions based on the elevation and the spectral indices such as NDVI and nDBR and (2) analyze the linear relationship of biomass regeneration and examine above-ground biomass (AGB) by observing the relationship between the WUI and non-WUI region and the level of fire severity such as low and moderate fire severity.

Methods

Study Area

This study was conducted in the San Gabriel Mountains of Los Angeles County, focusing on a tract that extends from 34°18'49" to 34°19'34" North and 118°8'14" to 117°56'14" West (figure 1). The Station Fire was one of the largest wildfires in California that combusted 160,557 acres of land. The Station Fire was first recognized on the 25th of August 2009, and was not contained until almost two months later on the 16th of October. Prior to the wildfire, chaparral was the most dominant vegetation type in this region, covering 75% of the total surface of the perimeter. Other vegetation types such as conifers, oaks, and other hardwoods, as well as desert shrub were also present in the study area. Within this part of the Angeles National Forest, slopes ranged between 20% and 60%, with elevations ranging from 397 to 2432 meters above sea level (m asl) (Thompson et al., 2021). Prior to the Station Fire, the vegetation of the area did not experience any significant fire activity in the previous 40 years, which created a prominent fuel source for when the 2009 fire began. Moreover, the absence of significant winds and drought stress exacerbated the vulnerability of the fire. The perimeter of the fire is shown in figure 1 (Holtzclaw, 2021).

Field Methods and Data Acquisition

The study focused on elevation variations and on spectral indices such as the Normalized Difference Vegetation Index (NDVI) and Normalized Difference Burn Ratio (nDBR). Visual predictor models were created to evaluate the initial hypothesis that the expanding WUI region and/or human disturbances have negatively influenced post-wildfire regeneration. Figure 1 demonstrates how WUI areas have expanded into the study region along the northern and the southwestern peripheries of the Station Fire perimeter, plus near the Big Tujunga Creek in the southwest portion of the study area.

Figure 1

Location, study area, burn perimeter, and wildland-urban interface of the 2009 Station Fire.

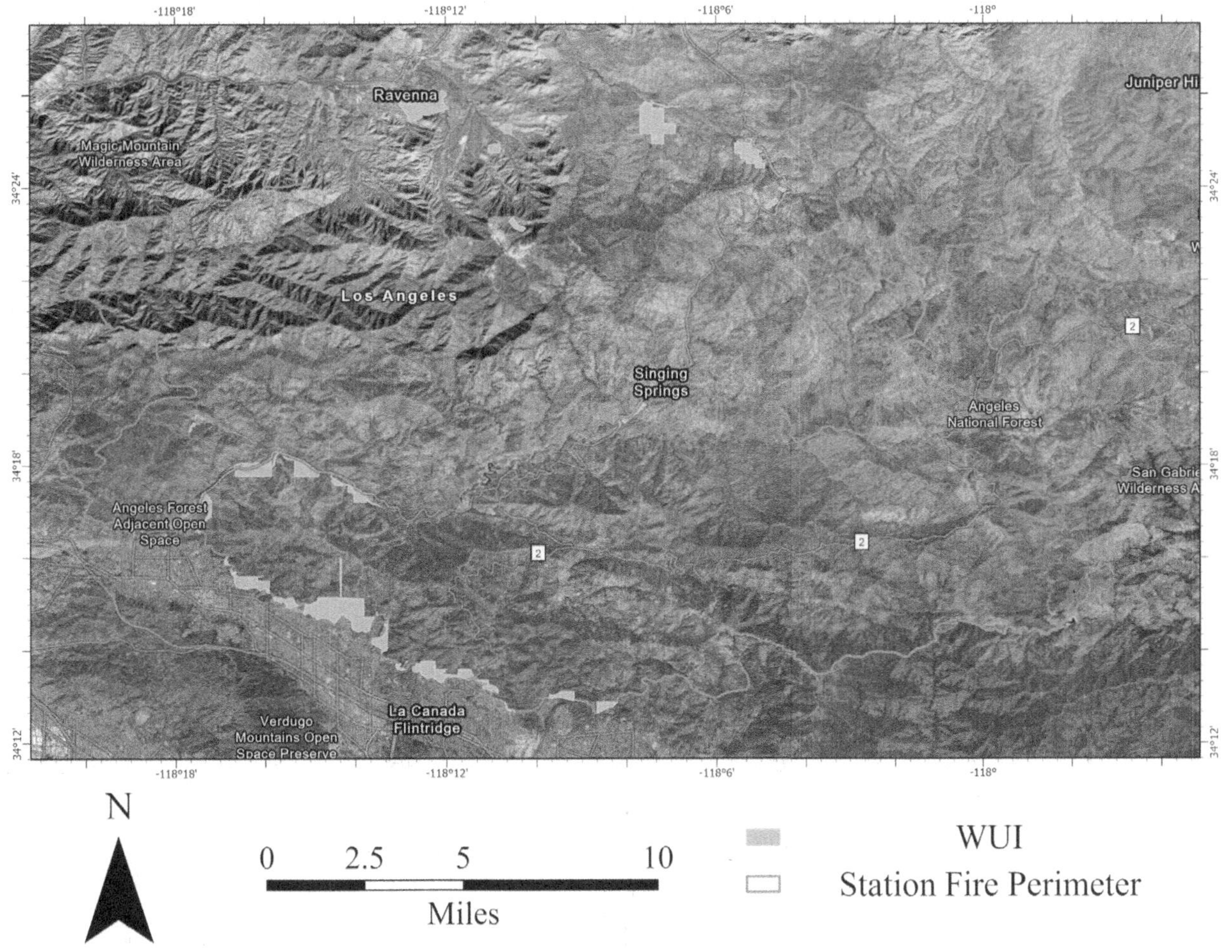

Topography and Hydrology

Elevation data were obtained from the Shuttle Radar Topography Mission (SRTM) Digital Elevation Model (90 m, spatial resolution) dataset (Peterman, 2013). Major rivers and streams were retrieved from the National Hydrology Dataset (n.d.). Historical fire perimeters were acquired from the California Department of Forestry and Fire Protection's Fire and Resource Assessment Program (FRAP) Fire Perimeters dataset (FRAP, 2021).

Wildland-Urban Interface (WUI)

The WUI data were acquired from the Spatial Analysis for Conservation and Sustainability (SILVIS) Lab at the University of Wisconsin-Madison (Gilbert, 2019). The WUI was defined as the area of intermix where man-made structures are built near or among lands prone to wildfire. The expansion of the WUI region directly introduces more human activity into areas that were formerly characterized as pure wildland. It is important to note that the expanded WUI does not include any attendant increases in responsible environmental preservation or conservation policies, which ended up decreasing the amount of defensible space and increasing the spread of wildfires. The urban sprawl in WUI areas has also led to the increase in wildfire prevalence in California.

Normalized Difference Vegetation Index (NDVI)

The NDVI is a technique to measure health, greenness, and vegetation abundance. NDVI pre-fire and post-fire conditions were used to calculate the greenness of the dominant vegetation. NDVI was calculated using Landsat 5 Thematic Mapper (TM+) images obtained from the United State Geological Survey (USGS). The satellite images retrieved for the study were captured on July 7 and October 25, 2009. The NDVI index ranged from -1 to 1. Large negative values indicated unhealthy vegetation, sparse vegetation, or dead vegetation. Large positive values indicated abundant, green, and healthy vegetation. Sometimes, obstructive clouds, snow, ice, or water can interfere with NDVI values. In some cases, an NDVI value of 0 revealed inorganic objects in the WUI such as human-made structures. To calculate the NDVI, formula 1 (below) was used, which takes near-infrared (NIR) and visible red bands, and divides them by their sum (Levin, 2019):

Formula 1:

$$\text{NDVI} = \frac{\text{NIR - RED}}{\text{NIR + RED}}$$

Normalized Difference Burn Ratio (nDBR)

Pre-fire and post-fire images were used to calculate the nDBR for the overall study area. The nDBR data were acquired from the USGS Landsat 5 TM+ imagery from July 7 to October 25, 2009. The Difference Burn Ratio (DBR) (formula 2) was applied to the before and after fire satellite images to calculate the ratio between NIR and shortwave-infrared wavelengths (SWIR) of the electromagnetic spectrum. After calculating the DBR, the nDBR was measured by the ratio between the pre-fire DBR and post-DBR (formula 3). Remotely sensed images were paired before and after the fire to determine the loss of vegetation biomass caused by fire. Char depth, reduced infiltration, loss of organic matter, changes in structure and phenotypic color, and the effects of fire on the ground surface were considered (Eidenshink et al., 2007).

Formula 2:

$$\text{DBR} = \frac{\text{(NIR - SWIR)}}{\text{(NIR + SWIR)}}$$

Formula 3:

$$\text{nDBR} = \text{PreFireDBR} - \text{PostFireDBR}$$

Large negative values indicated post-fire regrowth, whereas large positive values signified strongly burned areas with no post-fire regrowth. To define the nDBR values extracted as seen in formula 2 and formula 3, fire severity categories including "unchanged," "low fire severity," "moderate fire severity," and "high fire severity," were derived from a Composite Burn Index (CBI) (Miller & Thode, 2007). The relative index was chosen because of the heterogeneous composition of vegetation in the Station Fire perimeter. The thresholds of the relative index had greater accuracies than ecosystems with homogenous compositions. CBI was calculated by assigning values to the fire effects on the vegetation strata. During the evaluation process, fire effects such as char height, species mortality, and soil and composition changes were also taken into account, as well as vegetation strata like substrate, herbs, shrubs, and trees. After assigning threshold values to burn severity categories, the severity levels were determined.

Light Detection and Ranging (LiDAR) Data

LiDAR data in point cloud format were obtained from two free online databases: 2009 LiDAR data from Open Topography and 2016 LiDAR data from the USGS. LiDAR point returns were classified into the following categories: Unclassified, Ground, Low Vegetation, Medium Vegetation, High Vegetation, Building, Low Points, and Reserved, by using the Gaussian smoothing kernel to accurately find the apex of the vegetation for high frequency spatial filtering. The height or vertical filtering classified LiDAR points based on feature extractions using elevation, intensity, and echo times of the returned pulses. LiDAR data were processed using the vertical accuracy assessment to evaluate vegetated and non-vegetated areas.

Above-Ground Biomass

The model sampled 80 plots measured at 30 x 30 m2 across the Station Fire Burn Perimeter. Every plot in the WUI and non-WUI regions was approximated for the above-ground biomass (AGB) using the given LiDAR data. Above-ground biomass was defined as all living material above the Earth's surface layer expressed as kg/m2. The AGB was quantified using a random selection of predictors and samples from training data to measure ground truth biomass (Schrader-Patton & Underwood, 2021). The Digital Terrain Model (DTM), Digital Surface Model (DSM), and Canopy Height Model (CSM) were extracted from the LiDAR data. The DTM represented the bare ground and delineated the watersheds in the plot. The DSM analyzed the top of the vegetation canopy. The CSM was the difference between the DSM and CSM that provided the canopy height. Then, it was converted by the allometric equations from the Random Forest model to compute the AGB. The biomass allometric equation was applied to obtain a ground truth biomass as seen in formula 4. Where, B is biomass (kg/m2), H is height (m), C is cover (%), and BD is bulk density (kg/m3).

Formula 4: $$B = H \times C \times BD$$

The Random Forest Algorithm was implemented with the predictor biomass values to extract the ground truth biomass values to improve the accuracy. To avoid overestimation or underestimation of biomass, the forest regression model was used to accurately calculate the vegetation biomass in the WUI and non-WUI regions from the predictors. The output was the total above-ground biomass given in kilograms for the vegetation cover area sample plots.

Statistical Analysis

To assess if WUI regions contained different means of above-ground biomass (AGB) than non-WUI regions, a two-sample independent t-test was conducted for the 2009 and 2016 LiDAR data. Twenty plot samples for each region were tested to differentiate the AGB means (n = 40). Plots were chosen using convenience sampling to conduct respective analyses. Simple linear regression models were performed to quantify vegetation recovery rates in WUI and non-WUI regions. The study sampled 20 plots for each region to obtain the linear regression equation. The degree of fire severity and the type of region were also analyzed using a two-way ANOVA (2 x 2 factorial design) to see if there was a significant difference in AGB means between these two variables, and to determine if fire severity did not impact the type of region. The study sampled 10 plots for each condition: WUI and moderate fire severity, WUI and low fire severity, non-WUI and moderate fire severity, and non-WUI and low fire severity (n = 40).

Results

The digital elevation models demonstrated that elevation impacted vegetation structure, composition, and recovery in the WUI and non-WUI regions. The Station Fire elevation map displayed WUI areas in the Northern regions (600 m asl on average), Southwestern regions (500 m asl), and adjacent to the Big Tujunga Rivers (500 m asl) (figure 2). Within the Station Fire perimeter, there was a clear spatial relationship between the WUI regions and elevation. WUIs were more likely to occur in lower elevations and flatter topography.

Compared to WUI areas, non-WUI regions were spread across a wider range of elevations from 397 m asl to 2432 m asl. Non-WUI areas were dominated by natural vegetation without the presence of roads, houses, and buildings. On another note, some non-WUI regions were near major rivers and streams such as the Pacoima Wash, the West Fork San Gabriel River, the Santa Clara River, and the Mill Creek (figure 2). Therefore, non-WUI regions have ideal growth conditions for natural vegetation and are typical of chaparral-dominated landscapes.

Normalized Difference Vegetation Index (NDVI)

In general, NDVI displayed significant differences in the WUI regions than in the non-WUI regions of the pre-wildfire study area models. On July 7, 2009, there was an abundance of vegetation. Spatially, WUI regions displayed positive and negative NDVI index values equivalent to healthy and unhealthy vegetation (figure 3). Northern WUI regions captured smaller plots of dispersed green vegetation. In contrast, Southwestern WUI regions and WUI regions near the bodies of water displayed a larger spatial arrangement of greener vegetation. Some of the WUI regions showed neutral NDVI values representing urbanized areas such as roads and houses.

Pre-wildfire non-WUI regions displayed mixed results based on the elevations. Non-WUI regions in higher elevations revealed unhealthier vegetation than in lower elevations. Based on the topography, land regime, and degree of fire intensity, the non-WUI regions were more vulnerable and combustible at higher elevations than at lower elevations. Higher elevations led to faster fire progression due to the heat rising in the air. The fire ignitions at high elevations demonstrated a distinct relationship between higher elevations and landscape. As a result, the pre-wildfire WUI regions had comparable spatial arrangements of green vegetation to the pre-wildfire non-WUI regions.

Post-wildfire, the NDVI values were overall neutral and represent dead vegetation (figure 3). WUI and non-WUI regions recorded similar spatial patterns of non-productive and unhealthy vegetation. Within the WUI, there were WUI defining characteristics such as houses and buildings.

Normalized Difference Burn Ratio (nDBR)

The nDBR of the Station Fire measured a change in biomass levels by calculating the difference between the pre-wildfire and the post-wildfire image. The Station Fire burn severity analyzed the regions of different burn severities post-wildfire. Figure 4 indicated WUI areas were primarily defined as low fire severity. Northern WUI regions displayed low fire severity and moderate fire severity. Similarly, Southwestern WUI regions consisted of unburned and low fire severity. Unburned nDBR negative values denoted insignificant changes in predicted vegetation regrowth. Low fire severity measured small, positive nDBR values, which predicted a slower vegetation regrowth period. Moderate fire severity quantified large, positive nDBR values, indicating the slowest vegetation regrowth period.

On the other hand, non-WUI areas displayed a range of nDBR values. The non-WUI areas generally consisted of unburned, low fire severity, and moderate fire severity in high and low elevations. The eastern non-WUI areas showcased a dominant unburned area. High elevations suggested moderate fire severity and low elevations suggested low fire severity. The non-WUI did not have a concrete rebound arrangement when compared to the WUI. Therefore, the vegetation regeneration pattern in the non-WUI varied differently than the WUI.

Figure 2
Digital Elevation Model (DEM) of the 2009 Station Fire showing elevation variation.

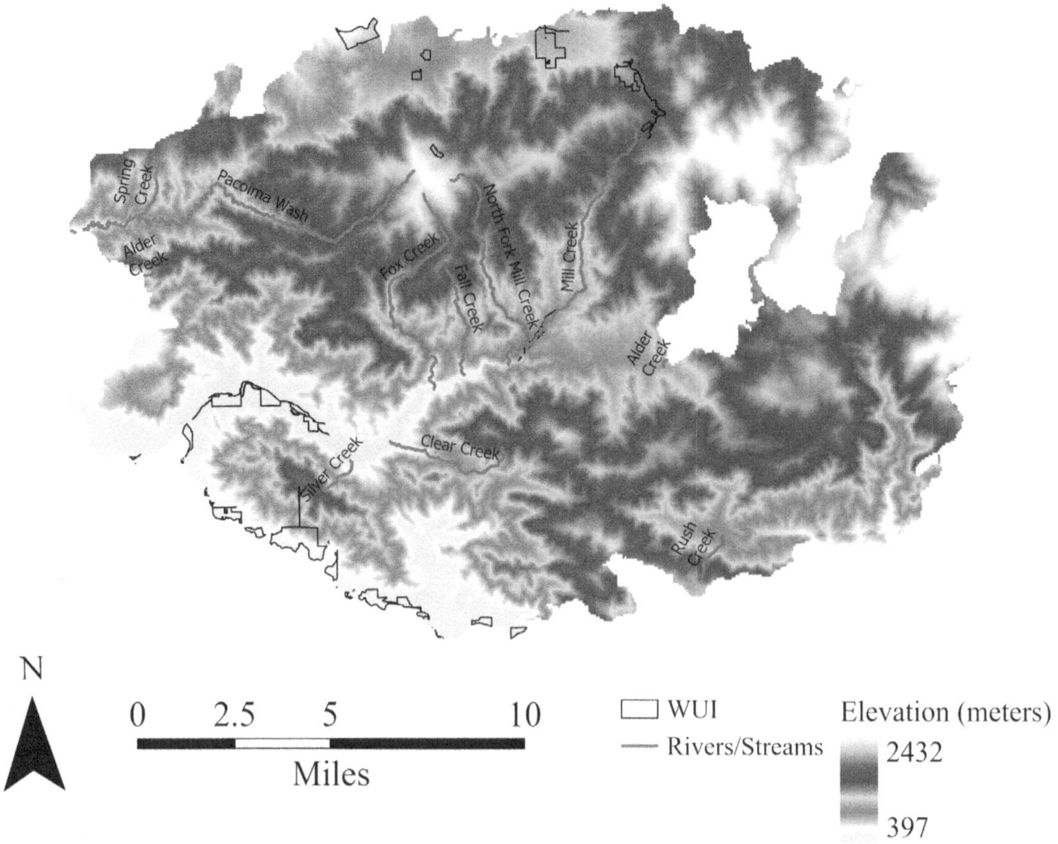

Figure 3
Pre-wildfire NDVI , July 7, 2009; Post-wildfire NDVI, October 25, 2009 of the Station Fire.

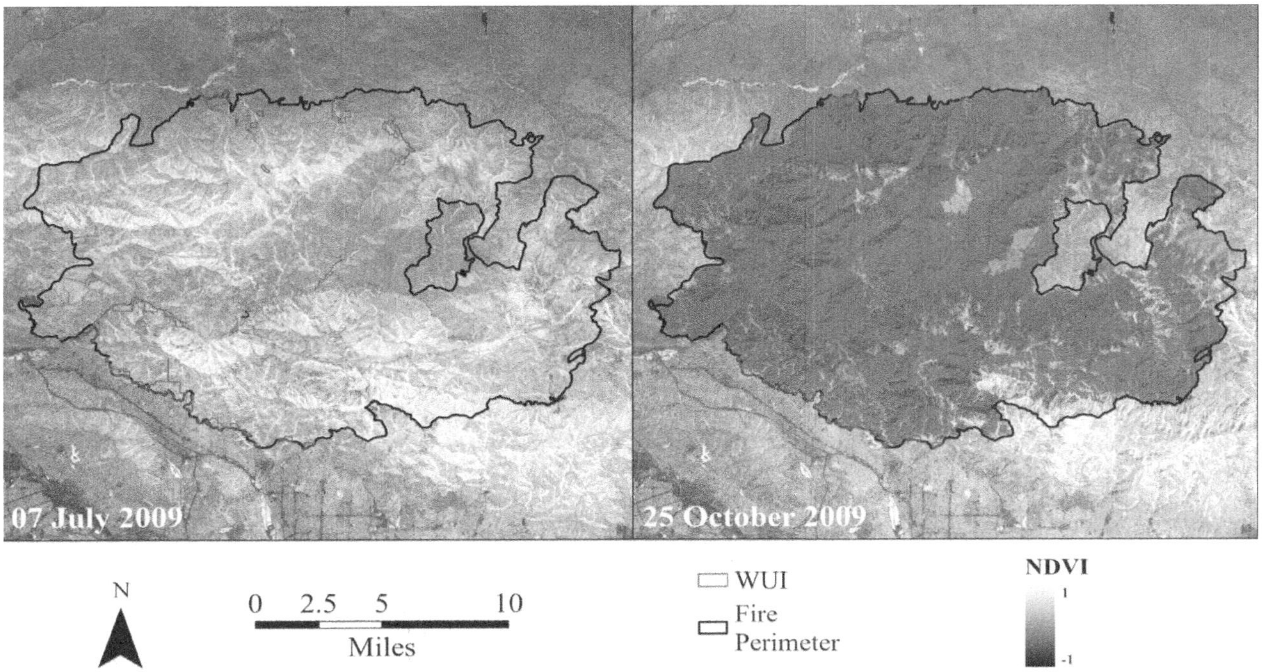

Figure 4
Normalized Difference Burn Ratio (nDBR) of the Station Fire. Unchanged, low fire severity, and moderate fire severity were listed to measure future regeneration.

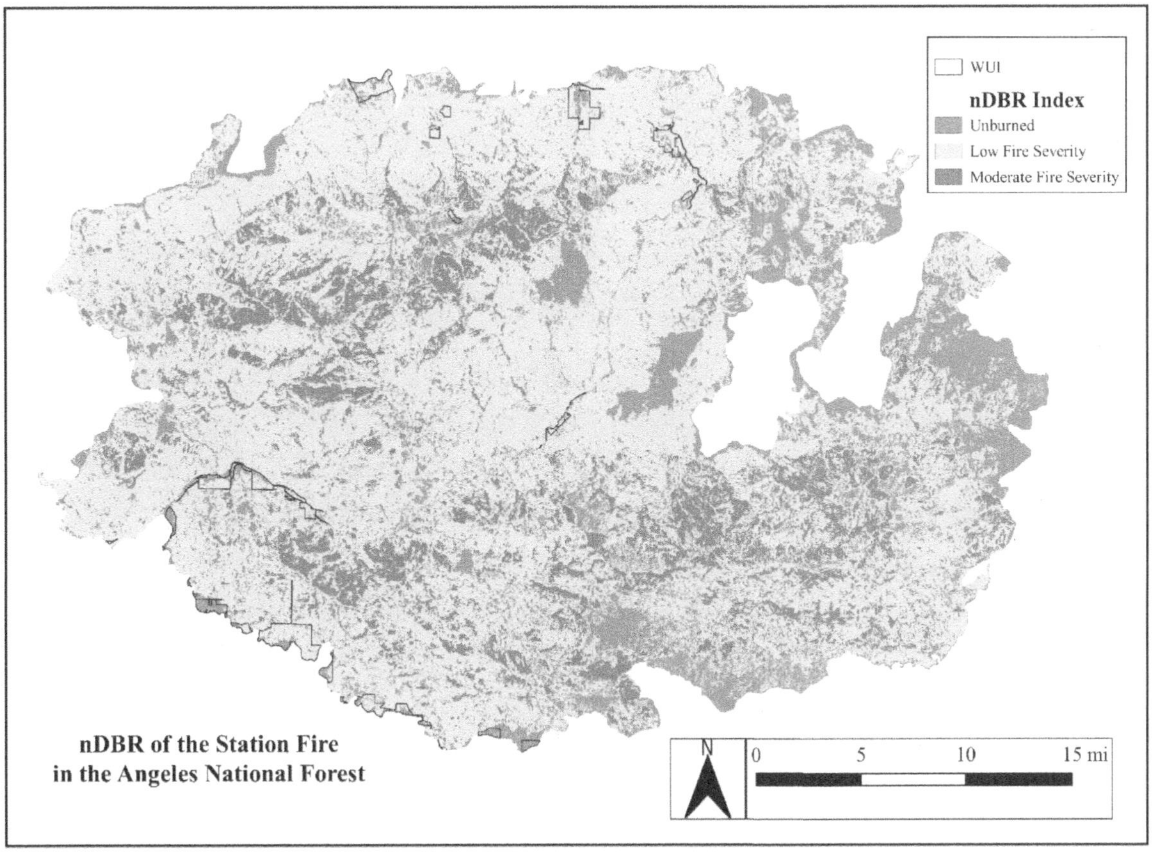

LiDAR Above-Ground Biomass Analyses

There was no significant difference in 2009 above-ground biomass between WUI (M = 10,989.31, SD = 2,098.99) and non-WUI (M = 10,163.75, SD = 747.14) regions; t (38) = -1.66, p = 0.10. The 2016 analysis also did not show a significant difference in biomass between WUI (M = 19,9075.40, SD = 119,454.98) and non-WUI (M = 232,784.30, SD = 36,866.54) regions; t (38) = 1.21, p = 0.23 (figure 5).

Two simple linear regression models were conducted to compare vegetation recovery rates for WUI and non-WUI regions shown in figure 6. Simple linear regression models tested if non-WUI regions significantly predicted AGB. The overall regression was statistically significant ($R^2 = 0.95$, $F_{1,38} = 729$, $p < 0.01$). For WUI regions, too, there was a statistically significant regression ($R^2 = 0.57$, $F_{1,38} = 49.57$, $p < 0.01$).

The two-way ANOVA analysis showed a significant interaction between the type of region and fire severity (F1, 36 = 14.54, p <= 0.01) (figure 7). Simple main effects analysis showed that the type of region did not have a statistically significant effect on post-wildfire ABG regrowth (p = 0.13). Simple main effects analysis yielded a statistically significant impact on the level of fire severity and post-wildfire ABG regrowth (p = 0.01).

Discussion

The study hypothesized that aboveground biomass would recover differently post-wildfire based on the NDVI, nDBR, and the level of fire severity and the type of region. Visual accounts of NDVI suggested the consumption of the majority of the vegetation post-fire. There was uniformity of the vegetation biomass pre-fire and post-fire when comparing WUI and non-WUI regions. The similar spatial litter of biomass in both regions showed that the vegetation biomass was resilient to human disturbances and different levels of fire severities. Seven years after the Station Fire, the biomass regenerated quickly, demonstrating similar seed dormancy patterns. In the presence of humans, the overall vegetation thrived in fire-prone environments.

Visual accounts of nDBR predicted specifically that WUI regions would contain less abundant AGB than

Figure 5

The difference in 2009 and 2016 above-ground biomass for WUI and non-WUI regions. The figure on the left demonstrates the similar mean chaparral AGB for the WUI and non-WUI regions post-wildfire in 2009. The figure on the right demonstrates the similar mean chaparral AGB for the WUI and non-WUI regions seven years after the fire in 2016. There was an outlier for 2009 and 2016 post-wildfire and seven years after the fire.

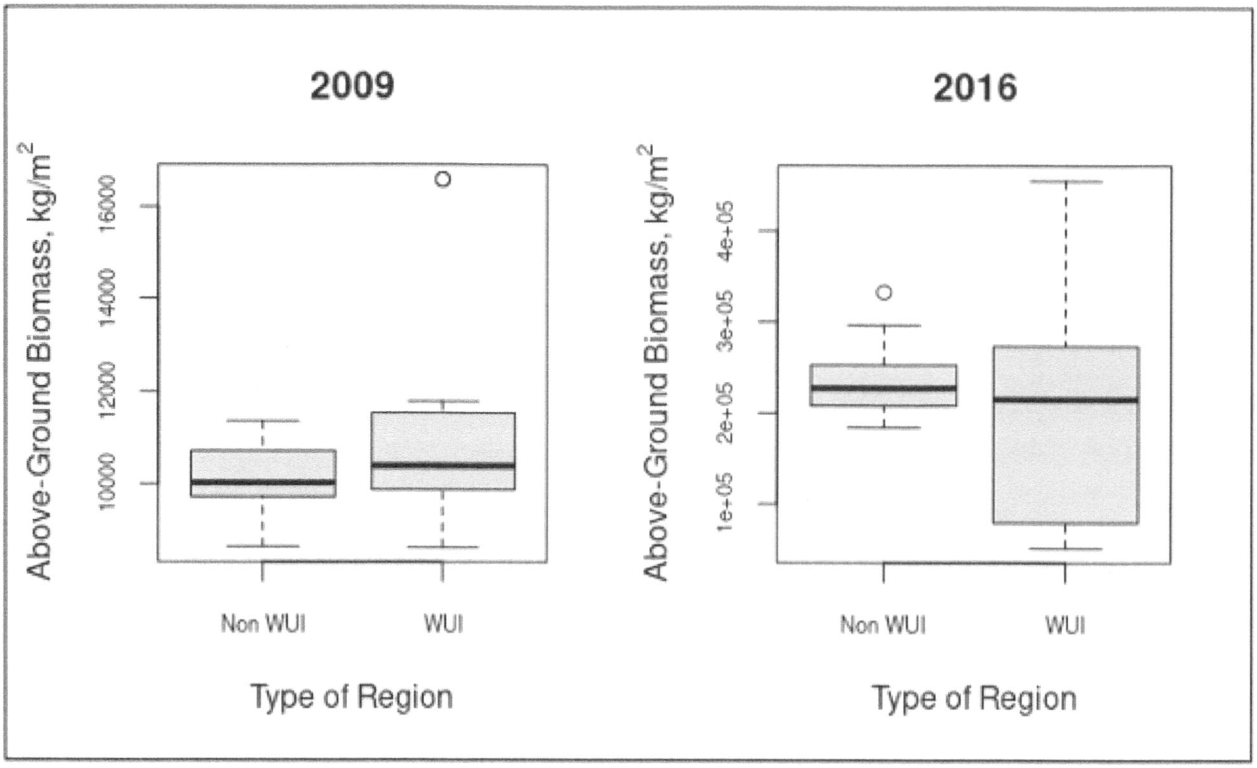

Figure 6

Vegetation recovery rates for WUI and non-WUI regions. The figures (from left to right) depict the individual plots acquired from 2009 and 2016 for the WUI and non-WUI regions, respectively. The blue line demonstrates the linear regression line and slope to quantify the chaparral AGB recovery from 2009 to 2016. Non-WUI regions exhibited a faster rebound slope than the WUI region.

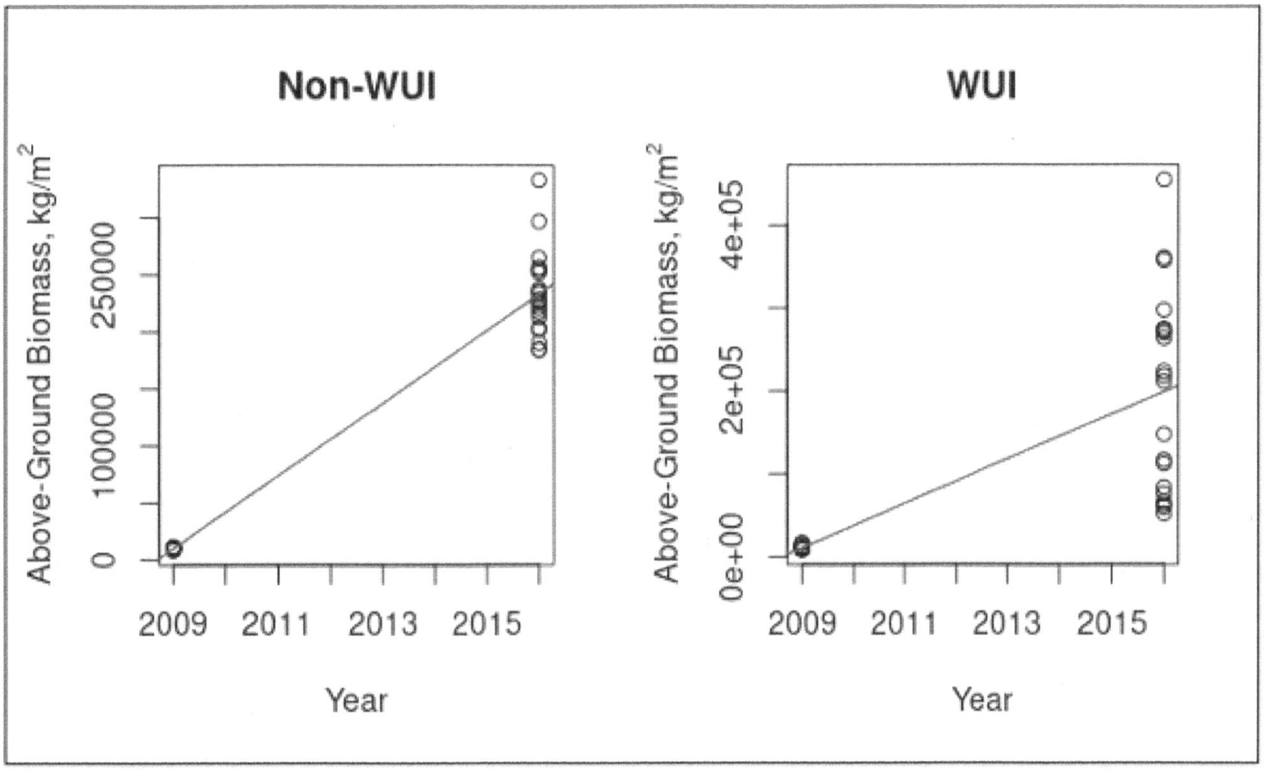

Figure 7

Interaction between the type of region and the type of fire severity level. The mean chaparral AGB for 2016 data was quantified by analyzing the severity and type of region of fire. This resulted in an interaction between the two independent variables. The mean chaparral AGB were similar for the type of region and the mean AGB for the level of fire severity were significantly different.

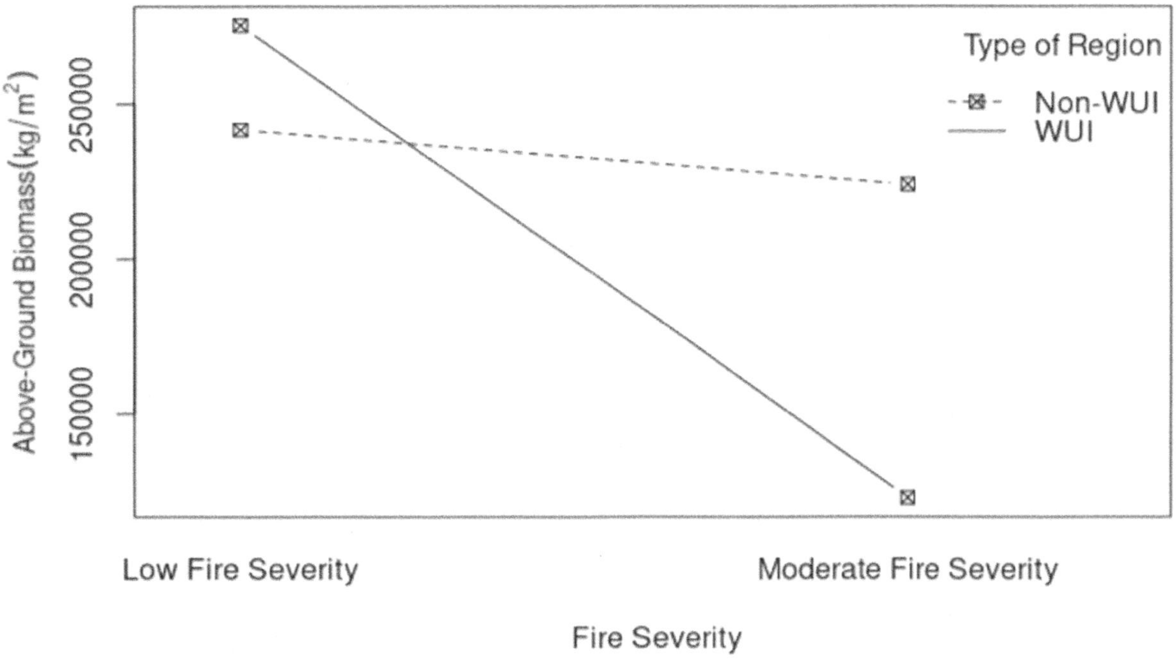

non-WUI regions due to heightened human activity. This hypothesis was not supported by the analyses, which meant that there were other contributing factors as to why vegetation biomass was abundant in both WUI and non-WUI regions. Elevation was one of the visual models that aligned with the original hypothesis. Though WUI regions were in lower elevations, the vegetation biomass was able to regenerate quicker than in higher elevations. In the lower elevations, there was low fire severity. The crown cover retained its structure and regrew faster than in rural areas. Because higher elevations consisted of moderate fire severity in the non-WUI region, the ecosystem was able to adapt and succeed in these fire prone environments quickly. Based on the linear regression analyses, the vegetation recovery period was comparable for WUI and non-WUI in 2009 through 2016. Although the rates of recovery were relatively similar, the vegetation biomass in the WUI was lower than in the non-WUI. This slight distinction should not be overlooked because it indicates that urbanization plays a minor role in vegetation recovery. The similarity was also attributed to how robust the chaparral ecosystem is ubiquitous to the environment. The vegetation structure of the chaparral post-fire was more easily converted into other vegetation types, usually non-native grassland, when it experienced several changes to its fire regime.

The mapping of aboveground biomass was based on fire severity and the type of region associated with higher fire risks in WUI than in non-WUI regions. Based on the given data, WUI expansion might have an influence in exacerbating the vulnerability of the regions to higher wildfire intensities and combustion. The level of fire severity depended on the type of region. It should be noted that overall, AGB did not elicit significant differences between WUI and non-WUI regions. The general classes (low fire severity and moderate fire severity) were associated with the region (non-WUI and WUI). Other factors such as temperature, microclimates, and precipitation may serve as reasons for an equal vegetation rebound in WUI and non-WUI regions.

Some limitations of the study were attributed to the Landsat 5 TM+ satellite images. With a temporal resolution of sixteen days, Landsat 5 TM+ monitored a small range of areas. A close temporal relationship between the satellite images before and after the fire was required for the NDVI and nDBR spectral indices. Errors in the data scanning process can occur due to the timeframe of the satellite images. In addition to the spatial resolution, there was another limitation to the study. Because Landsat 5 TM+ has a spatial resolution of 30 m, the pixelation of images did not capture detailed information. Higher resolution images will increase precision and accurately define vegetation biomass.

Another limitation of the study was the acquisition of the temporal data. Two years of data were analyzed in the study, but the results are not clearly indicative of fire recovery. There is a possibility that vegetation in the WUI could have recovered more quickly and then slowed down, resulting in a lower average biomass in the WUI. Data access limitations made it difficult to accurately depict recovery patterns.

A major concentration of chaparral vegetation was found in the Angeles National Forest following the Station Fire. It exhibited a range of phenotypic colors, from green to brown. Several types of chaparral vegetation can be identified by their differences in coloration. There is also the possibility that the WUIs contain non-native tree and shrub species cultivated for landscaping, and which are regularly watered, which might influence the NDVI of the WUIs and potentially affect the severity index. The NDVI could have underestimated healthy vegetation in the WUI regions. In contrast to NDVI, nDBR measures vegetation recovery based on fire severity. It is possible to misinterpret post-fire images as arid and dry areas due to drought instead of a wildfire.

Conclusion

Comparing vegetation biomass in the WUI and non-WUI regions post-wildfire of the Station Fire in the Angeles National Forest allows us to better understand vegetation regrowth in the presence of human disturbances. The land use and land cover type in rural areas also play a significant role in vegetation rebound. First, the vegetation biomass responded similarly in WUI and non-WUI regions in both 2009 and 2016. Second, the vegetation linear succession was analogous in WUI and non-WUI regions. Third, the level of fire severity yielded a significant effect on the post-fire biomass regrowth from 2009 to 2016. While this is the result of the uniqueness of the chaparral vegetation, it reflects how the chaparral can survive short-interval and long-interval fires in conjunction with human disturbances. The chaparral is well-adapted to recurring burning; however, the wildfire prevalence can create dissonance for the future based on higher fire severities. The study revealed new information about the correlation between high elevations and moderate levels of fire severity. Fires at higher elevations caused more destruction by damaging the vegetation composition and increasing the susceptibility to crown fires. As a result, the fire patterns at higher elevations removed the plant crown cover and reduced seed dormancy. It is also important to note that the chaparral community will thin with the repeated burning of these areas. Moderate severities that occur on slopes can increase the susceptibility to surface runoff, influence the repellency of the soil, and reduce the vegetation biomass in these elevations. Moderate fire severities are fatal for subsequent wildfires. Moderate fire severity and high elevations have the tendency to reburn post-wildfire. While it is possible for the chaparral to be well-adapted and resilient to any extreme climatic and anthropogenic factors, human disturbances may play a minimal role in vegetation rebound. To better understand the implications of these results, future research should consider examining water quality, slope, and aspect, to address the similarities of recovery rates in WUI and non-WUI regions. Carbon sequestration should also be assessed to measure plant productivity.

References

Barro, S. C., & Conard, S. G. (1991). Fire effects on California chaparral systems: an overview. *Environment International, 17*(2), 135–149. https://doi.org/10.1016/0160-4120(91)90096-9

California Department of Forestry and Fire Protection's Fire and Resource Assessment Program (FRAP, 2021). Post-fire soil erosion. Retrieved September 15, 2021, from https://frap.fire.ca.gov/mapping/gis-data/

Efthimiou, N., Psomiadis, E., & Panagos, P. (2020). Fire severity and soil erosion susceptibility mapping using multi-temporal Earth Observation data: The case of Mati fatal wildfire in Eastern Attica, Greece. *CATENA*, 187, Article 104320. https://doi.org/10.1016/j.catena.2019.104320

Eidenshink, J., Schwind, B., Brewer, K., Zhu, Z.-L., Quayle, B., & Howard, S. (2007). A project for monitoring trends in burn severity. *Fire Ecology, 3*(1), 3–21. https://doi.org/10.4996/fireecology.0301003

Gilbert, M. (2019). Wildland Urban Interface in CA. Retrieved September 15, 2021, from https://services1.arcgis.com/pf6KDbd8NVL1IUHa/arcgis/rest/services/Wildland_Urban_Interface_vector/FeatureServer

Guo, X., Coops, N. C., Tompalski, P., Nielsen, S. E., Bater, C. W., & John Stadt, J. (2017). Regional mapping of vegetation structure for biodiversity monitoring using airborne lidar data. *Ecological Informatics*, 38, 50–61. https://doi.org/10.1016/j.ecoinf.2017.01.005

Holtzclaw, J. (2021). California Fire Perimeters 1878–2020. Retrieved September 15, 2021, from https://services.arcgis.com/jIL9msH9OI208GCb/arcgis/rest/services/California_Fire_Perimeters_1878_2019/FeatureServer

Keeley, J. & Safford, H. (2016). THREE. Fire as an Ecosystem Process. In H. Mooney & E. Zavaleta (Ed.), *Ecosystems of California* (pp. 27-46). Berkeley: University of California Press. https://doi.org/10.1525/9780520962170-007

Levin, D. (2019). Vegetation recovery after wildfire in the Santa Monica mountains: A remote sensing analysis [Bachelor's thesis, University of Pennsylvania]. https://www.researchgate.net/publication/340874625_Vegetation_Recovery_After_Wildfire_in_the_Santa_Monica_Mountains_A_Remote_Sensing_Analysis

Martín-Alcón, S., Coll, L., De Cáceres, M., Guitart, L., Cabré, M., Just, A., & González Olabarría, J. R. (2015). Combining aerial LiDAR and multispectral imagery to assess postfire regeneration types in a Mediterranean forest. *Canadian Journal of Forest Research*, 45(7), 856–866. https://doi.org/10.1139/cjfr-2014-0430

McCarley, T. R., Kolden, C. A., Vaillant, N. M., Hudak, A. T., Smith, A. M., Wing, B. M., Kellogg, B. S., & Kreitler, J. (2017). Multi-temporal LiDAR and Landsat quantification of fire-induced changes to forest structure. *Remote Sensing of Environment,* 191, 419–432. https://doi.org/10.1016/j.rse.2016.12.022

Miller, J. D., & Thode, A. E. (2007). Quantifying burn severity in a heterogeneous landscape with a relative version of the delta Normalized Burn Ratio (dNBR). *Remote Sensing of Environment, 109*(1), 66–80. https://doi.org/10.1016/j.rse.2006.12.006

National hydrography (n.d.). National Hydrography Dataset. Retrieved November 7, 2021, from https://www.usgs.gov/core-science-systems/ngp/national-hydrography/national-hydrography-dataset?qt-science_support_page_related_con=0#qt-science_support_page_related_con.

Peterman, W. (2013). 90 m DEM of California, USA. Retrieved October 20, 2021, from http://databasin.org/datasets/78ac54fabd594db5a39f6629514752c0

Rull, V., González-Sampériz, P., Corella, J.P., Morellón, M., & Giralt, S. (2011). Vegetation changes in the southern Pyrenean flank during the last millennium in relation to climate and human activities: the Montcortès lacustrine record. *Journal of Paleolimnology*, 46(3), 387–404. https://doi.org/10.1007/s10933-010-9444-2

Schrader-Patton, C. C., & Underwood, E. C. (2021). New Biomass Estimates for Chaparral-Dominated Southern California Landscapes. *Remote Sensing (Basel, Switzerland)*, 13(8), 1581–. https://doi.org/10.3390/rs13081581

Storey, E. A., Stow, D. A., O'Leary, J. F., Davis, F. W., & Roberts, D. A. (2021). Does short-interval fire inhibit postfire recovery of chaparral across southern California? *The Science of the Total Environment, 751*, 142271–142271. https://doi.org/10.1016/j.scitotenv.2020.142271

Syphard, A. D., Rustigian-Romsos, H., & Keeley, J. (2021). Multiple-scale relationships between vegetation, the wildland–urban interface, and structure loss to wildfire in California. Fire, 4(1), 12. https://doi.org/10.3390/fire4010012

Tepley, A. J., Thomann, E., Veblen, T. T., Perry, G. L. W., Holz, A., Paritsis, J., Kitzberger, T., & Anderson-Teixeira, K. J. (2018). Influences of fire-vegetation feedbacks and post-fire recovery rates on forest landscape vulnerability to altered fire regimes. *Journal of Ecology,* 106(5), 1925–1940. https://doi.org/10.1111/1365-2745.12950

Thompson, R., Kaplan, C., Gomberg, D. (2021). The Station Fire: An example of a large wildfire in the absence of significant winds. Retrieved October 20, 2021, from https://www.weather.gov/media/wrh/online_publications/talite/talite1002-1.pdf

Wulder, M., White, J., Alvarez, F., Han, T., Rogan, J., & Hawkes, B. (2009). Characterizing boreal forest wildfire with multi-temporal Landsat and LIDAR data. *Remote Sensing of Environment,* 113(7), 1540–1555. https://doi.org/10.1016/j.rse.2009.03.004

Made in the USA
Middletown, DE
05 November 2026

36544977R10038